Weather and Climate

Weather and Climate

An Introduction to Meteorology

Edited by
ROGER SEARS

With a foreword by
PATRICK MOORE

ORBIS PUBLISHING · London

A

Contents

Frontispiece: Lightning over Waskesiu Lake, Saskatchewan (Colour Library International)
Endpapers: Namib Desert, Namibia (Carol Hughes/Bruce Coleman)

© Istituto Geografico de Agostini, Novara 1974
English edition © Orbis Publishing Limited, London 1974
This edition published in 1978
Originally published as The World's Weather and Climates
Printed in Italy by IGDA, Novara
ISBN 0 85613 148 2

Foreword

The weather is always with us. It is something which affects all our lives, and yet something which we cannot control to any marked extent – and it makes up a large part of our everyday conversation. Surely there can be nobody who does not, at some time of the day, read a weather forecast in the newspaper or watch the weather report on television. Very often these forecasts are accurate; not infrequently they are wrong, and the luckless holidaymaker who has been drenched to the skin after leaving his raincoat at home may have some hard things to say about meteorologists in general.

Criticism of this kind is easy to make, but it is somewhat unfair. Until recent times we really knew very little about the earth's atmosphere, except that it is limited in extent and that it is in a state of constant turmoil. Today the situation has been transformed. We have new techniques, and also new means of exploration; for the first time we can study the atmosphere 'from above', by using satellites and space-probes.

There is also the question of climate. It is too easy to say 'Summers aren't what they used to be forty years ago!', but it is possible that there may be more in this than meets the eye. Climate does alter, both over short-term and long-term periods; and this again is linked with the structure and constitution of the earth's atmosphere.

Meteorology, together with climatology, is a fascinating science, and much more involved than some people think. This book gives a clear account of what weather study is all about, using some remarkably fine illustrations to drive home the points which are made. It could not have been written a decade ago; we know so much more now than we did even in the early 1960s, and so its appearance is timely. The scope is wide, covering the atmosphere itself, the lofty regions high above the earth, the structure of clouds, weather systems, climatic changes and, of course, that topic which is so much in the news today: atmospheric pollution.

There are, I think, many people who look at the synoptic charts shown on television several times per day, and who have no real idea of what the maps or the relevant terms mean. Anyone who follows the present book through will end up with a good understanding of what meteorology is all about – after which the synoptic charts will start to make sense. The reader will also begin to realize how fascinating meteorology and climate studies really are. I am glad to recommend the book to you.

PATRICK MOORE

Glossary

Anticyclone High-pressure system.

Convergence Type of motion in fluids (such as air) which involves a general flow inwards to a point from all directions. Such a flow can only be maintained by a vertical flow from the point of convergence.

Cyclone Low-pressure system.

Divergence Type of motion in fluids which involves a general flow outwards from a point in all directions. Such a flow can only be maintained by a vertical flow to the point of divergence.

Front Boundary between air streams which differ in such properties as their temperature or their moisture content. This boundary takes the form of a sloping surface extending from the ground and sloping towards the colder and more dense of the two air streams. Warm fronts, cold fronts and occluded fronts are the types most frequently encountered. At a warm front, warm air is advancing and replacing colder air; at a cold front, cold air replaces warm air. An occluded front is associated with a wedge of warm air which, in a well developed cyclone, has been forced upwards so that it is no longer in contact with the ground. The occluded front may have the characteristics of either a warm or a cold front. Weather changes are to be expected at and near fronts, generally involving some form of precipitation.

Ionosphere Region of the atmosphere at altitudes of roughly 80 km to 400 km (50 to 250 miles) where the concentration of ions and electrons is most dense. The ionization is caused by the interactions of gamma rays, X-rays and short-wavelength ultraviolet rays present in solar radiation with the molecules and atoms of nitrogen and oxygen in the upper atmosphere.

Jet Stream Band of very strong winds (very often in excess of 100 mph) which can be between 150 km and 450 km (90 and 280 miles) in width. It is best visualized as an oval-shaped tube (the larger dimension being the horizontal one) through which air flows very rapidly. The 'tube' itself may be more than 1500 km (930 miles) long, and often describes a sinuous or wavy path. It is itself carried along, relatively slowly, in the atmosphere. Jet streams have been reported at many levels in the atmosphere including short ones quite close to the ground. The most important and familiar ones, however, are those of the upper troposphere at about 12 km ($7\frac{1}{2}$ miles).

Kelvin Temperature Scale (°K) This scale, commonly used in scientific research, has the same degree units as the Celsius (Centigrade) scale. Its zero point is at what is known as 'absolute zero', approximately $-273°C$. Thus $0°C$ is equivalent to $273°K$; $100°C$ is equivalent to $373°K$, and so on, up the scale.

Mesopause Boundary between the mesosphere and the thermosphere. Found at about 80 km (50 miles) height, it is characterized by a change from decreasing temperature with height in the mesosphere to increasing temperature with height in the thermosphere.

Mesosphere Upper layer of the atmosphere extending from 50 km to 80 km (31 to 50 miles). Within this layer temperature decreases with increasing height. In this region the shortest wavelengths of solar radiation react with oxygen molecules, bringing about ozone formation.

Millibar Unit of air-pressure measurement used in meteorology. One millibar is equivalent to 1000 dynes/square cm of surface. Pressure distribution can be shown by drawing **isobars,** lines joining places with the same air-pressure values in millibars (mbs) over the area in question.

Radiation Energy transmitted in a vibratory or wave-like motion, at a constant speed of 186,000 miles per second (the 'speed' of light). The amount and the wavelengths of the radiation emitted by a body depend upon its temperature. Bodies like the sun which are at very high temperatures radiate more energy and at shorter wavelengths than bodies at low temperatures, like the earth.

Stratopause Boundary between the stratosphere and the mesosphere at a height of about 50 km (31 miles). It can be identified as the point where temperature ceases to increase with height as it does in the mesosphere.

Stratosphere Atmospheric layer extending in altitude from about 13 km to 50 km (8 to 31 miles). In the lower portion of this layer temperature is nearly constant with height but, in the upper portion, temperature increases with increasing height. Most of the atmosphere's ozone is contained in the stratosphere.

Thermosphere Uppermost layer of the atmosphere extending upward from 80 km (50 miles). It is a region where temperatures increase rapidly with height but also where there are large diurnal fluctuations of temperature. Atomic oxygen and atomic nitrogen become important constituents of the thermosphere above a height of about 100 km (62 miles).

Tropopause Boundary between the troposphere and the stratosphere. It occurs where the decrease of temperature with increasing height (characteristic of the troposphere) gives way to approximately **isothermal** (constant temperature) conditions in the lower stratosphere. On the average the tropopause slopes from the equator where it is at a height of 18 km (11 miles), to the poles where it is at 8 km (5 miles).

Troposphere Lowest layer of the atmosphere extending from the surface to a height of about 13 km (8 miles). In this layer temperature decreases with height at a rate of roughly 7°C (13°F) per kilometre. This layer contains most of the mass of the atmosphere and most of the important weather systems.

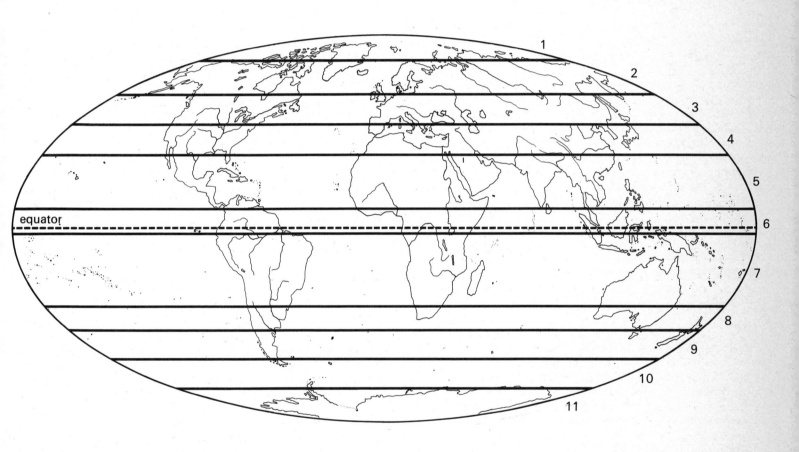

Diagram 1

Latitudinal zones of surface winds and pressure

The map above shows how the major features of atmospheric movement and pressure are related to latitude and to the geography of the world. These features are the basis for the distribution of weather systems and hence for the phenomena of climate. The greater area of the inter-tropical convergence zone to the north of the equator is due to the far greater areas of land in the Northern Hemisphere

Key to Diagram 1
1   polar easterlies
2   marine low pressure
3   westerlies
4   subtropical highs
5   trade winds
6   intertropical convergence zone
7   trade winds
8   subtropical highs
9   westerlies
10 low pressure
11 polar easterlies

List of Plates

EARTH IN SPACE

If you were an observer looking towards the Milky Way galaxy, across an unimaginable distance from some other galaxy, you would find it difficult to count the stars making up the disc-shaped mass with its spherical nucleus. An estimate of the number of stars in the Milky Way would be of the order of billions; perhaps 200,000,000,000 stars might be a reasonable figure. You would note also, that the stars which made up this spiral galaxy are not all the same size, the biggest stars being hundreds of thousands of times larger than the smaller ones. Further, their temperatures differ so that the hottest stars are about a hundred times hotter than the coolest. With all the wonders of this galaxy would you, an observer from intergalactic space, bother to look closely at a rather cool, average-sized star about one third of the way out from the nucleus of the galaxy and very close to its central plane? That star is our sun and one of its smaller planets is the earth. Although it is a rather ordinary star, the sun governs the past, present and future for life on earth.

Both archaeological evidence and written historical records show how real was the bond man felt with the sun. The sun was a god in ancient Egypt and in pre-Columbian South America both Aztec and Inca societies worshipped the sun. To these and many other societies our star was a god whose warmth and light were needed for crops to grow; a god who gave the seasons which were the pulse of human existence, which measured the passage of time and the growth of history and identity. Today man is no less dependent on the sun than he was thousands of years ago. The growth of scientific knowledge has left no room for a deity drawn across the sky in a fiery chariot, but the star remains to radiate heat and light, maintaining the environment on the surface of the planet earth. So it is with the sun and its radiation that our account of the weather and the atmosphere must begin.

For most of us there is little need to understand processes of stellar birth and death, indeed relatively few people have any idea how our star produces its energy. The source region for the radiant energy which we can see, feel and measure is in the interior depths of the huge ball of gas some 864,000 miles in diameter. Deep within the star, where enormous pressure gives gases a density about ten times that of lead and temperatures are probably as high as 15,000,000 °K, nuclear fusion reactions take place. Here, in the sun's core, various fusion reactions are likely to be involved in the energy generating process, but the simplest one is the so-called proton-proton reaction whereby four hydrogen nuclei combine to form one helium nucleus. The combined mass of four hydrogen nuclei is slightly larger than that of one helium nucleus, so there is an excess mass which is converted to energy in the reaction and released. The scale of this fusion reaction is so immense that the sun is losing mass at the rate of 4.5×10^{12} g per second. This is, however, such an insignificant proportion of the total mass of hydrogen available that mankind need have no fears that our star may die in the foreseeable future.

The energy released from the proton-proton reaction in the interior is transported outwards as very short wavelength radiation such as gamma rays, but after travelling through most of the solar mass the radiation is absorbed in an outer region. The huge amounts of energy absorbed here set up vast, turbulent convection currents which carry the energy outwards to that part of the sun which we see as a bright disc – the photosphere. The photosphere represents the top of this convective zone where the hot gases lose energy by radiating out to space, thus cooling, and then sinking towards the sun's interior again. Very powerful solar telescopes and particularly balloon-borne and satellite-mounted telescopes show the photosphere as possessing a 'granulated' surface somewhat similar in appearance to the pattern you can see on the surface of a cup of hot tea or coffee. The brighter 'granules' are in fact the tops of the hot, rising gas cells which will cool and sink back into the darker background, to be replaced by other bright, hot cells. This photospheric region has significantly lower pressure, density and temperature than the interior so that the energy radiated from the outer edge of this zone is of longer wavelengths than the energy radiated from the helium core.

Above the photosphere are two regions of the sun, the chromosphere and the corona. These two very turbulent zones can be seen with the naked eye only during a total eclipse. Some absorption of the photosphere's energy takes place by lower temperature gases (4,300 °K) at the interface between the photosphere and the chromosphere. The density of the gases in these outer regions of the sun drops off very rapidly with increasing

(Above) Andromeda is a spiral galaxy, probably quite similar to our own galaxy the 'Milky Way'

(Above right) the sun. In the interior, nuclear fusion reactions produce enormous quantities of energy which are ultimately radiated into space

(Left) chromosphere, one of the outer layers of the sun, can only be seen with the naked eye when the photospheric disc is covered, as during a total eclipse. At such a time prominences (vast 'fountains' of ionized gas) can be seen

(Right) the solar disc photographed in the red light emitted by hydrogen atoms. The bright, hotter areas are called plages and represent solar disturbances. Dark filaments on the disc represent cooler prominences; sun spots are seen as black, rounded structures

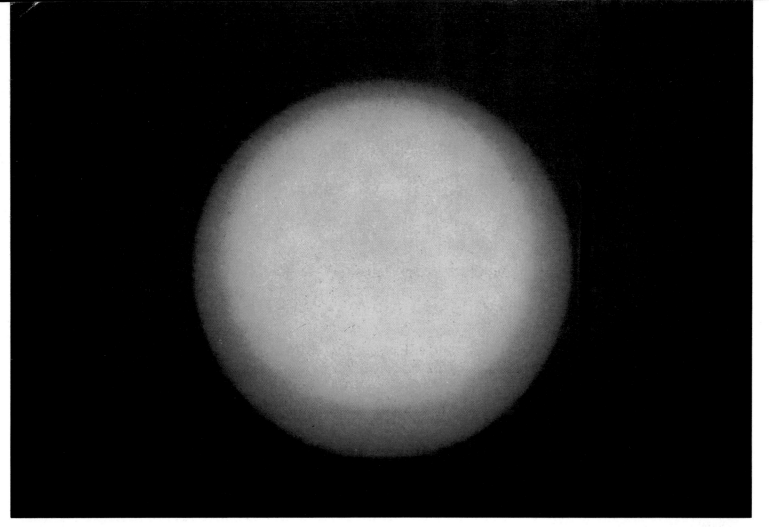

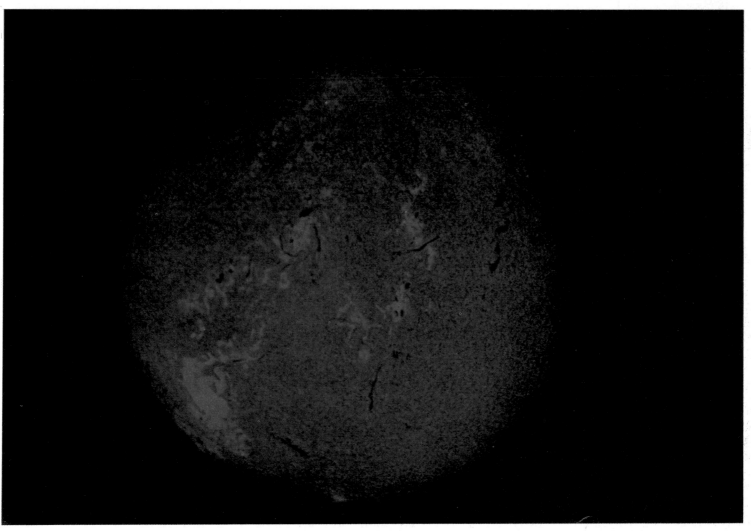

3

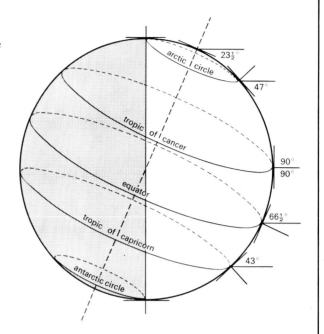

Diagram 2: the orientation and orbit of the earth
The intensity of solar energy received at a point on the earth's surface is largely determined by the sun's angle of elevation. This angle, as can be seen (right) is determined by the tilt of the earth's axis of rotation. The diagram shows the Northern Hemisphere's summer solstice when the tilt of the earth's axis produces maximum solar elevation angles at noon to the north of the equator, and thus gives the Northern Hemisphere the greatest intensity of solar energy

arctic circle
$23\frac{1}{2}°$
$47°$
tropic of cancer
$90°$
$90°$
equator
$66\frac{1}{2}°$
tropic of capricorn
$43°$
antarctic circle

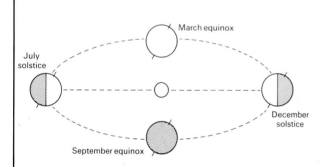

July solstice

March equinox

December solstice

September equinox

The seasons and the weather owe their existence to the characteristics of the earth's motion around the sun, especially to the fact that the earth's axis of rotation has a constant tilt as it travels around the sun with an elliptical path. This gives each hemisphere one period each year when it is tilted toward the sun (the summer season) and one period when it is tilted slightly away from the sun (the winter season). The transitional periods, when neither hemisphere is so favoured, represent spring and autumn

distance from the photosphere 'surface', but temperature increases from some 30,000 °K in the chromosphere to 2×10^6 °K in the tenuous gas of the corona. The corona extends outwards so for that the earth (and the inner planets) could be said to be in orbit within the sun's upper atmosphere, even though the earth is, on average, 93,000,000 miles from the sun!

To all intents and purposes, the rate at which the sun pours out energy into space is constant when it is examined in terms of a human time scale. If one imagines a large sphere enclosing the sun, but at earth's mean orbit distance from it (i.e. 93×10^6 miles), then solar energy would be passing outwards through the sphere at an approximate rate of 5.8×10^{27} calories per minute (or about 3.9×10^{21} kw). Just a tiny fraction of this outpouring maintains the environment of our planet. Despite the virtually constant energy supply provided by our star, the organization of life on earth is almost everywhere adapted to seasonal change and subtle short-term variety of living conditions. The explanation has to be sought by looking not just at the earth or the sun but at the relationship between them.

The earth revolves around the sun not with a

truly circular orbit – planets never do – but with an elliptical path. A complete revolution takes 365·24 days, marginally longer than a calendar year. The plane of the path traced out by the revolving earth is called the ecliptic. The earth also rotates, spinning about its own polar axis of rotation to complete one rotation with respect to the sun about every 24 hours. But the axis of rotation through the poles and the centre of the earth is not perpendicular to the ecliptic. It is tilted $23\frac{1}{2}$ degrees from such a perpendicular and pointing in a fixed direction with respect to the stars. Thus, as the earth revolves around the sun, for a certain part of the orbit the axis of rotation will be tilted so that the Northern Hemisphere will be pointed slightly towards the sun and the Southern Hemisphere slightly away from it. Some six months later the positions will be reversed when the earth has travelled half way around its orbit. Obviously there will be two periods of transition when neither hemisphere will be favoured by being tilted towards the sun. The seasons are therefore produced by the earth revolving around the sun with a tilted but constant axis of rotation (see diagram 2). At about 21 June, when the Northern Hemisphere has its

(Right) this magnificent view of the earth and its atmospheric weather patterns was taken by the Apollo 11 *crew. The continents, oceans, ice caps and cloud systems present a picture of striking beauty. The African continent dominates the photograph, whose coverage stretches from the Arctic to the Antarctic regions*

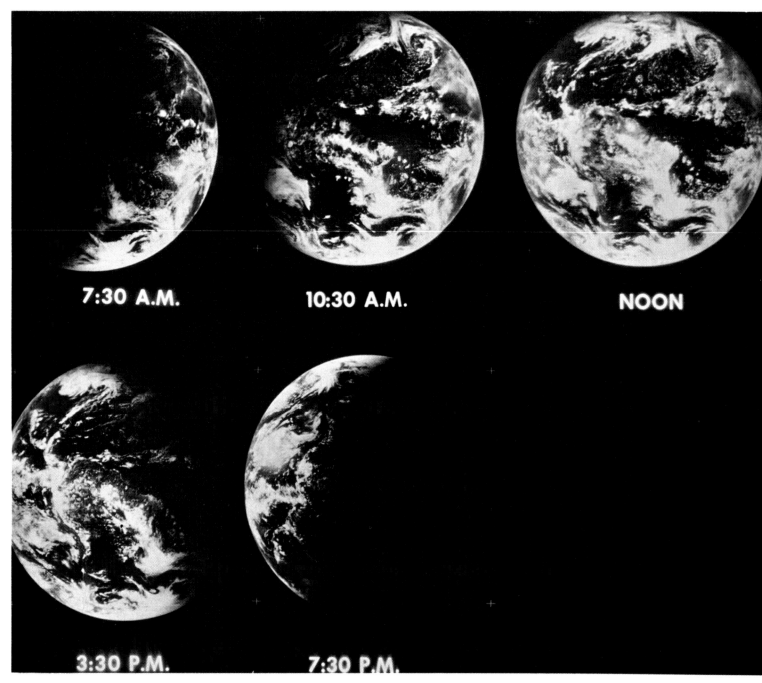

7:30 A.M.   **10:30 A.M.**   **NOON**

3:30 P.M.   **7:30 P.M.**

(Above) a satellite stationed high above the South Atlantic just off South America recorded this series of pictures in November 1967. The Western Hemisphere and the satellite revolved into the sunlight just after the first picture was taken; the third picture shows the whole earth seen by the satellite and lit by the sun's rays. In the last two pictures the satellite sees the Western Hemisphere revolving into darkness. Since the pictures were taken in late November the Northern Hemisphere is tilted away from the sun and receives less illumination

(Opposite page) sunsets can be both beautiful and dramatic. When the sun is just above the horizon its light follows the longest path through the atmosphere. This means that more light scattering can take place. Since this scattering is most effective at the blue end of the visible light spectrum, only the orange and red portion of the spectrum reach the ground directly

maximum tilt towards the sun, the sun's rays are perpendicular to the earth's surface at the Tropic of Cancer ($23\frac{1}{2}$ °N. latitude) where the heating of the surface will be most intense. This gives the Northern Hemisphere its summer while the Southern Hemisphere experiences its winter since the sun's rays strike the surface more obliquely there. On about 22 December, when the tilt is reversed, the Southern Hemisphere experiences its summer solstice. The vernal and autumnal equinoxes, 21 March and 22 September respectively, occur when the tilt of rotation favours neither hemisphere.

Locally the rotation of the earth superimposes a diurnal rhythm of heating and cooling on the seasonal change, but one must look to the nature and composition of the atmosphere to see how more complex variations, known as weather, arise.

(Left) a rainbow is seen when you stand with your back to the sun looking at a rain-shower. The bows or arches are formed by the raindrops reflecting and refracting the sunlight. The primary bow can be quite intense with violet on the concave side, red on the convex side. The secondary bow, caused by more than one internal reflection of sunlight in the rain-drops, has the sequence reversed and is usually rather faint

(Right) this 15th century illustration, drawn from the book Splendor solis, *shows reverence for the sun a. the bringer of heat and light*

THE VERTICAL ORGANIZATION OF THE ATMOSPHERE

As the sun is eclipsed by the black disc of the earth the thin shell of gases which constitute the atmosphere shows as a narrow, bright arc above the planet's surface. If the earth were the size of an apple the total thickness of the atmosphere would be about the same as the skin of the fruit. (Photograph taken by Apollo 12 astronauts)

In comparison with the vast proportions of the solar 'atmosphere', the thin skin of gases on which life on earth depends is of trivial dimensions. This is dramatically seen on some of the photographs taken from spacecraft, which, looking down at the obliquely illuminated horizon, show the atmosphere as only a thin, bright line above the surface of the planet. However small its dimensions on a planetary scale, the atmosphere is fundamental to survival and its characteristics are of great importance in an explanation of weather phenomena.

Before trying to determine what happens to the solar energy stream as it penetrates the atmosphere, one should be aware of the composition of this gaseous mixture. If for the moment water vapour is excluded, then it is found that dry air has an almost uniform composition to heights of at least 160 km (100 miles). Although many gases are present in the mixture, the greatest proportions by volume are contributed by nitrogen (N_2) with 78 per cent, and oxygen (O_2) with nearly 21 per cent. The next most common gas is argon (A) at 0.9 per cent followed by carbon dioxide (CO_2) at 0.03 per cent, which means that all other gases including helium, hydrogen and so on, together account for less than 0.07 per cent by volume of dry air. But perfectly dry air is never encountered; for water is always detected mixed with the atmospheric gases in varying proportions (these variations being both in time and space) up to perhaps 4 per cent by volume. Even though water vapour and other constituents make up such a small proportion of the atmospheric volume, life has developed on earth in such a way that they are of crucial importance. The carbon content of the biosphere comes from carbon dioxide, and water vapour profoundly affects the climatic environ-

ment in many ways besides providing the water for rain and snow. Another constituent of the atmosphere which cannot be neglected is particulate matter. Although the solid particles have to be very small in order to remain in suspension for long periods, they play an important role in the atmospheric processes, notably those connected with precipitation (the formation of rain drops or snow crystals). It is impossible to list all the substances which may be present as particulate matter and the distribution is variable, but in the lower atmosphere, salt crystals from the oceans, dust from the deserts, and vegetable matter such as pollen and spores are common. Very fine particles are also found at some distance from the surface; one such layer where sulphate particles are common is found at a height of 18 km (11 miles). More recently manned spacecraft have brought back evidence of another dust layer which may be as high as 40 km (25 miles).

Through this atmosphere the solar energy must pass before it can be received and perhaps utilized at the surface of the earth, but during its passage certain very important effects take place. The solar beam approaching the earth is in the form of electromagnetic radiation transmitted through space like radio waves, but because of the very high temperature of its photospheric source (6,000 °K) the wavelengths are very short. Most of the energy, amounting to about 2 calories per square centimetre per minute (1.4 kw per square metre) is found between 0.15 μm and 4.0 μm (1 μm, i.e. 1 micrometre, equals one millionth of a metre). Some emissions of much shorter and much longer wavelengths are present but contribute very little to the energy of the radiation stream. The first interactions which occur between the incoming solar stream and the earth take place at distances up to 64,000 km

(40,000 miles) away from the planet. Here the earth's magnetic field begins to trap high energy electrons (part of the solar stream) which are then channelled down towards the earth's magnetic poles. On some occasions they excite the gases as they enter the upper atmosphere in high latitudes, to produce auroral glow at heights up to 250 km (155 miles) by a mechanism similar to that of fluorescent lamps. The major part of the solar energy stream continues through these regions unaffected until at heights of about 1,000 km (620 miles) it begins to enter the upper, tenuous part of the earth's atmosphere where atomic oxygen is the largest constituent gas followed by atomic nitrogen. By what may be thought of as a collision process between these gas atoms and the shortest wavelengths of the energy stream, electrons are stripped away, leaving the gas ionized (positively electrically charged) and the energy stream depleted of the gamma rays, X-rays and some of the ultra-violet rays – the wavelengths involved in the collisions. Since density increases in the atmosphere as the earth's surface is approached (this being a simple expression of gravity acting on the gases) the collisions become more frequent until, at about 80 km (50 miles), virtually all the rays capable of ionization have been used up. Thus the *ionosphere*, the layer of ionized gas produced by this interaction of atmosphere and solar radiation, may be thought of as a filter cutting off some very harmful rays which would otherwise be hazardous to the life forms on the planet's surface.

The next part of the atmosphere to be penetrated by the incoming energy is a region known to meteorologists as the *mesosphere* at 50–80 km (35–50 miles). Here molecular nitrogen and molecular oxygen are the commonest gases but it is with the oxygen that the radiation tends to interact. The shorter wavelengths of the ultra-violet rays energize and break up oxygen molecules into oxygen atoms, some of which may move to higher levels in the atmosphere, while others may recombine into molecules or react with oxygen molecules to form ozone (O_3). Individual ozone molecules are not very long-lived but the production rate and the rate of decay (which depend on both radiation intensity and oxygen concentration), balance to give the greatest ozone concentrations lower down in a yet denser part of the atmosphere known as the *stratosphere* at 10–50 km (6–35 miles).

The radiation is now reduced to longer wavelength ultra-violet rays, visible light (0·4 μm–0·7 μm) and infra-red rays (0·7 μm–4·0 μm). It now has to pass through the most dense parts of the atmosphere before reaching the earth's surface.

(Left) when the sun is just below the horizon a satellite or space-craft view of the atmosphere not only reveals the very shallow 'skin' of air which supports life on the planet but also some of the beautiful colour bands due to selective absorption and scattering of the sun's radiation by gases and dust layers

(Right) salt crystals enter the air over the sea from spray droplets which subsequently evaporate. They are among the particles which have great importance in the condensation processes producing clouds and rain

In the stratosphere, air molecules and particles begin to scatter the shortest wavelengths (ultra-violet, violet and blue), and it is this scattering of the blue content of the solar beam which we see as blue sky. Scattering becomes more efficient in the lowest and most dense region of the atmosphere, the *troposphere*, which extends from the surface to about 10 km (6 miles). Here more frequent scattering brought about by larger molecules and particles which scatter over all the visible wavelengths, vary the sky colour from blue to very pale blue or even to milky-white, depending on the concentration of 'scatterers'. Since half the total mass of the atmosphere lies below the 5 km (3.5 miles) level, it is not difficult to understand why the sky appears to be a deeper blue from a high mountain or an air-craft. Also, from such a high altitude it is possible to see how efficiently lower-lying clouds reflect solar radiation. Thus, depending on cloud depth, the amount of energy reaching the ground is reduced very effectively. If the sky is clear of clouds then perhaps 80 per cent of the original solar energy stream reaches the surface of the earth, but if clouds are present it may be further reduced to less than 40 per cent of its original strength.

Even after reaching the ground some of the solar radiation is reflected. Those who enjoy spring skiing know how important it is to wear goggles or dark glasses to protect the eyes – not only from direct sunlight but also from the strong glare of the light reflected by the 'snowfields'. Forests and fields reflect much less but there is still a proportion of sunlight reflected even by these darker surfaces. What remains of the radiant energy is absorbed and goes to heat up the top layer of ground or to penetrate and warm a thicker layer of water if it strikes a lake or sea surface. Some of the energy is stored up in the tissue of growing plants; some is stored for a season in soil or marine surface waters. Much of the absorbed energy is used to evaporate water, a process in which plants play a part, and some returns to the atmosphere as the heated ground warms the overlying air. By far the greatest proportion of the originally absorbed solar energy is re-radiated by the earth's surface. Since the temperature of the earth's surface is only about 1/20th of the solar photospheric tempera-ture, the energy is re-radiated at much longer wavelengths (4 μm–40 μm) than solar energy, in fact wholly in infra-red rays.

An interesting point now arises, for the gases of the atmosphere permit only a small number of these wavelengths to penetrate into space, the rest are efficiently absorbed by water vapour, carbon dioxide and ozone.

This energy absorption warms the atmosphere which will then re-radiate the same infra-red rays. Of course the air radiates in all directions so that the ground receives back a proportion of the energy it originally received from the sun and subsequently radiated away. Thus, for the infra-red portion of radiation, the atmosphere acts rather like a blanket keeping the ground warmer – particularly at night when, without an atmosphere, temperatures would fall to perhaps −150°C (−240°F). The energy that the air re-radiates upwards eventually leaks out to space,

(Above) high-altitude observing stations provided the first information about the vertical structure of the atmosphere. Now that balloons, rockets and aircraft can be used, mountain-top observations tend to be more important for the measurement of radiant energy

(Above left) small, light and disposable instruments known as radiosondes provide much of the information a meteorologist needs about temperature, moisture and wind conditions in the layers of the atmosphere up to 30 km (20 miles). Carried by balloons filled with hydrogen or helium, a radio transmitter sends back the information while being tracked by radar. Here a radiosonde balloon is ready for launching from a meteorological station in the Antarctic

(Left) nacreous or 'mother of pearl' clouds are occasionally seen in the stratosphere at about 30 km (20 miles). They are formed when high mountain ranges displace deep currents of air upward. Air at high levels may thus be cooled sufficiently to condense its water vapour. In this picture, taken when the last rays of the setting sun illuminated the high cloud, the air has been displaced by the Norwegian mountains

Diagram 3: average temperature distribution with height in the middle latitudes
The rather complex temperature distribution is a result of radiative interaction and atmospheric motion generated by solar heating. In atmospheric studies it is convenient to divide the atmosphere into layers, these being identified by particular temperature criteria. For example, the troposphere, which contains most of the mass of the atmosphere, has a general decrease of temperature (a lapse) with increasing height

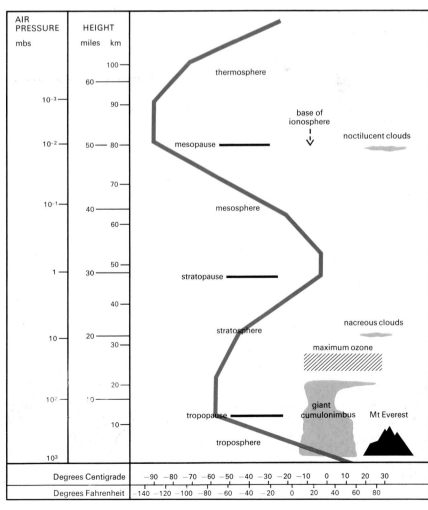

along with any infra-red rays which the atmosphere did not absorb. Over a period this energy loss to space balances the energy gained from the sun in such a way that the earth and its atmosphere grow neither appreciably hotter nor colder.

The interactions which take place between the atmospheric gases and radiant energy determine directly or indirectly the rather complicated relation between temperature and height in the atmosphere (see diagram 3 on page 17). Much of our knowledge of the average distribution of temperature with height has been acquired in recent years when the advance of technology has been most rapid. The lowest and most dense part of the atmosphere (the troposphere) shows temperature decreasing with height. Temperatures obtained from mountain-top observatories, balloons and aircraft confirm this upward decrease (or lapse) of temperature. In the stratosphere above 10 km (6 miles) temperatures at first vary little with height, but in the upper stratosphere radiation absorption by ozone and oxygen results in temperatures increasing steadily with height. Aircraft, balloons and rockets have supplied this information together with measurements of the amount of ozone at these heights. From 50 km (35 miles) upwards, in the mesosphere, temperature again decreases with height to very low values such as $-100°C$ ($-148°F$) at 80 km (50 miles). From here, the average temperatures begin once more to increase with height until the earth's upper atmosphere merges with that of the sun. Rockets have been used to verify this temperature structure at these high levels.

The most recent types of satellite have been designed to measure long-wave radiation passing up through the earth's atmosphere in such a way that the whole variation of temperature through the air below the satellite can be determined. This is obviously going to be a most useful technique since it will enable the average values to be checked over otherwise remote parts of the earth's surface. It will also allow meteorologists more insight into variations over shorter periods.

What we have described so far is the fundamental environment to which much of life is adapted. Life is shielded from harmful radiation by the atmosphere and protected from the extremes of temperature experienced by planets without appreciable atmospheres. Eyes are adapted to use the most intense wavelengths of solar radiation which reach the ground. In addition, the solar energy which in one form or another resides for a time within the earth and its atmosphere produces the weather upon which life is equally dependent but which occasionally becomes a hazard to survival.

THE TROPICAL ATMOSPHERE

The sun's rays are most efficient in heating a surface if that surface is at right-angles to the illuminating beam. The sun at noon produces greater warmth than the sun in the early morning simply because, in the latter case, the same cross-sectional area of a beam is spread over a larger area due to the oblique angle it makes with the receiving surface. In addition, the longer oblique path through the atmosphere will favour greater absorption, scattering and reflection, thus depleting the beam before it reaches the surface.

At the equator the noon-time position of the sun is never less than $66\frac{1}{2}$ degrees above the horizon, indeed twice a year (at the equinoxes) the noon-time sun is directly overhead. The poleward limit of the region where the overhead sun is experienced occurs at latitudes $23\frac{1}{2}°$N. and S., the parallels which are known as the Tropic of Cancer and the Tropic of Capricorn respectively. So it can be said that a belt of the earth's surface bounded by $23\frac{1}{2}°$N. and S. is favourably placed to be efficiently heated by solar radiation. It is no surprise then to find that on average this is the warmest region of the earth's surface.

Calculations have shown, however, that if one considers only solar radiational gains and infrared radiation losses, the intertropical zone should be warmer than it is, therefore some energy is being lost from the region rather than being retained to increase the temperatures. This loss takes the form of an export to middle and high latitudes which are warmer than radiation calculations alone would predict. It is the atmosphere and the oceans which act as export agencies. Air and water currents which carry away warm fluid (for air may be considered to be a fluid) and replace it with cooler fluid effectively export energy. It is exceedingly difficult to separate realistically the actions of air currents and ocean currents but it would appear that it is the atmosphere which carries out the larger part of this operation.

Large Scale Pressure and Wind Patterns

Air which is warmed by the ground becomes less dense and expands. On a local scale this warmed air may break away from the surface and rise through the surrounding air like a bubble rising through water. With large scale heating, great volumes of air can be expanded and since the expansion takes place upwards as well as outwards the thickness of the atmosphere is increased. At the same time the air pressure close to the ground surface decreases. Thus, in the equatorial region, we find average sea level pressure values are low, forming a belt or 'trough' of low pressure round the earth. It is also observed that the troposphere located over the equator is roughly twice as thick as it is over the poles of the earth.

With increasing distance poleward towards the tropics, pressure at sea level rises to quite high values so that each hemisphere is roughly girdled by high-pressure belts. These high-pressure regions are best developed over the tropical oceans of each hemisphere and are commonly known as subtropical high-pressure zones or anticyclones. Their pressure cannot be explained simply in terms of either radiation or average temperatures. Instead it is necessary to look at the average motions set up in the atmosphere above the intertropical zone by the abundant receipt of solar energy.

Imagine an air column over the equator being expanded by heating so that it protrudes up above the air columns some little distance poleward where heating is not so great. It may be likened to the tall office buildings of a city centre

(Left) meteorological sounding rocket being launched from White Sands Missile Range, New Mexico, U.S.A. Such rockets are used to determine the temperature and wind structure of the atmosphere at heights between 30 km and 80 km (20 miles and 50 miles)

19

protruding above the flats, warehouses and schools of the surroundings. At upper levels the expanded air column has a higher pressure than the unexpanded air. Under the influence of this pressure gradient air will flow out from the warmed region towards the cooler region but, since this represents a mass change, air will flow in at the ground and low levels to try to compensate for the upper-level loss. On one occasion it was estimated that more than 200,000,000 tons of air move towards the equator every second in the Northern Hemisphere alone. The low-level inflow will have to rise and the high-level outflow will have to fall in order that the atmosphere does not accumulate mass in any one place. So a circulatory air system (a circulation 'cell' of rising and falling air) is set up on a large scale. The descending branches of this cell are found at about 15°latitude to 25°latitude in each hemisphere. They produce the high-pressure belts, or anticyclones, while the ascending branch near the equator is identified with the surface trough of low pressure. Were it not for the fact that the earth is in rotation such a circulation might extend from equator to pole in each hemisphere. In the Northern Hemisphere this would give north winds everywhere close to the surface and south winds in the upper atmosphere but this pattern is modified in the following way.

Consider what happens when air at ground level at the equator begins to move up, then blow towards the north pole while the earth is rotating. By friction at the surface, the air has acquired some of the west-to-east motion of the earth and when it moves poleward it tends to keep this momentum but, as it travels north, it gets closer to the axis around which the earth rotates. Just like a spinning skater who spins faster when he pulls in his arms closer to his axis of rotation, so the air gains west-to-east speed as it moves poleward. So, instead of reaching the pole it ends up as an upper-level, high-speed, westerly wind. The core of this wind is found at a height of about 12 km (7·5 miles) above latitude 30°N. where it streams around the hemisphere with speeds in excess of 100 mph. This ribbon-like core of high winds is known as the subtropical jet stream.

At low levels the return flow from the sub-tropical anticyclones to the equatorial trough of low pressure is similarly affected by the rotating earth to give this air an east-to-west component of motion. This results in the wind belt known in the Northern Hemisphere as the NE. trades (the trade winds of the Southern Hemisphere are, of course, SE. winds). Thus the spinning earth confines the circulation cell to roughly the inter-tropical regions. The picture that has been put together so far explains only the broadest, most general features of the low-latitude atmosphere; namely the equatorial trough, the subtropical high pressures, easterly trade winds at the surface and high-level westerly winds with the subtropical jet stream. This general model gives an indication of how energy is moved by the atmosphere. Energy from the sun, received at the surface, is transformed into energy in the atmosphere which is moved out with the upper-level winds to the low-latitude margins of the middle latitudes. For this reason the low latitudes are sometimes called the 'fire-box' of the atmosphere, using an analogy with a heat-engine.

Water Vapour, Clouds and Rainfall

A large proportion of the earth's surface in the intertropical zone is covered by oceans with warm surface waters. The evaporation of water from the surface of the sea provides the atmosphere with a great deal of water vapour which has important effects upon the air. Thus it is necessary to look at the moisture cycle in the tropical and equatorial regions. The air of the subtropical anticyclones is dry because it originated high up in the troposphere and during descent became compressed and warm. The trade winds, originating in the anticyclones, are also dry and warm, but as they pass across the warm surface waters of the tropical oceans (on their journey to the equatorial trough) evaporation injects more and more water vapour into the air. This addition of water vapour reduces the density of the air and there is a tendency for localized convection currents to be set up in the moist air.

As this air from the surface regions penetrates upward into the cooler surroundings it begins to lose some of its heat. Furthermore, the atmosphere surrounding the rising convection current grows rapidly less dense with distance above the surface, since it is not so compressed by gravity. This causes the rising air that came from the surface regions to expand both outwards and upwards and, in doing so, to use up energy and therefore cool still further. Convection clouds form where the air has cooled to such an extent that it can no longer hold the water vapour which condenses out into tiny droplets (say 10 μm in diameter). Thus, the upward moving currents are made visible by the myriads of droplets produced in the cloud-forming process. Convection clouds have a heaped-up appearance and hence are termed *cumulus* or *cumuliform* (heap-shaped) clouds.

In equatorial areas the cumulus clouds tower up to heights of perhaps 7,000 m but nearer to

(*Above*) *the horizontal belt of cloud almost in the middle of this picture shows the position of the intertropical convergence zone. North and South America can be seen clearly on the right. The scattered clouds in the South Pacific Ocean off the west coast of South America are associated with the South Pacific subtropical high. The photograph was taken from an* Apollo *space capsule*

(Above) the crew of Apollo 9 *photographed this intense tropical disturbance over the Amazon Basin. Its violent vortex-like structure bears a similarity to a hurricane*

(Opposite page) cumulus clouds over the Caribbean. This aerial view shows scattered small cumulus clouds typical of the trade wind belts as well as a cluster of clouds in which convection is reaching much higher levels

(Pages 24–5) the tropical land surface of Mexico (Campeche) warms up quickly during the day and from its heated surface convection currents rise forming cumulus clouds. Over the lagoon and the open waters of the Gulf of Mexico solar heating is not so rapid and the atmosphere is almost cloud free

the tropics the clouds may have their tops at only 2,000 m, the tops having a flattened appearance. This latter type of cloud is known as trade wind cumulus, and its restricted vertical development is due to the widespread, gentle descent of air on poleward sides of the tropical circulation cell. The general increase of cumulus cloud heights towards the equator shows the build-up of atmospheric moisture along the path of the trades. This represents energy added to the atmosphere for, when the water vapour condenses this energy will be released as heat (the latent heat of vaporization amounting to about 600 calories to change one gram of water to water vapour). Since the evaporation took place from the ocean and condensation takes place in the air, this represents a significant transfer of energy from the oceans to the atmosphere.

In the tall, 'boiling' cumuliform clouds close to the equator, the latent heat released enables the convection currents to reach great heights and, in the course of this development, cloud droplets collide with each other and coalesce to form raindrops big enough (3 mm dia.) to fall through the rising air which first brought about condensation. The rainfall of the equatorial regions is very high (about 2,000 mm falling per year). By the time the water has been rained out of this atmosphere, the latent heat has been used in warming and expanding the air. This provides the energy which drives the low-latitude atmospheric circulation.

Recent Discoveries

Over the last decade the development of satellites able to transmit back to earth pictures taken from a variety of great heights has given a new knowledge of clouds and cloudiness over tropical ocean areas. It now seems likely that only certain clouds take an active part in this fundamental energy cycle. These are exceptionally tall cumulus clouds (known as 'hot towers') which penetrate to great heights because they move up through columns of air which have been moistened by recent cloud formation. Hot towers, when they occur, are usually found embedded in clusters of cumulus clouds which are produced where a disturbance in the trade winds has given rise to some local inflowing (or convergence) in the horizontal air currents. The convergent zone between the NE. and SE. trades (equatorial trough of low pressure) is obviously the most favoured area for disturbances of the wind régime and hence for the frequent generation of hot towers.

Satellite pictures have shown that on average this convergence zone (or trough) is found some 5° north of the equator. This displacement is the

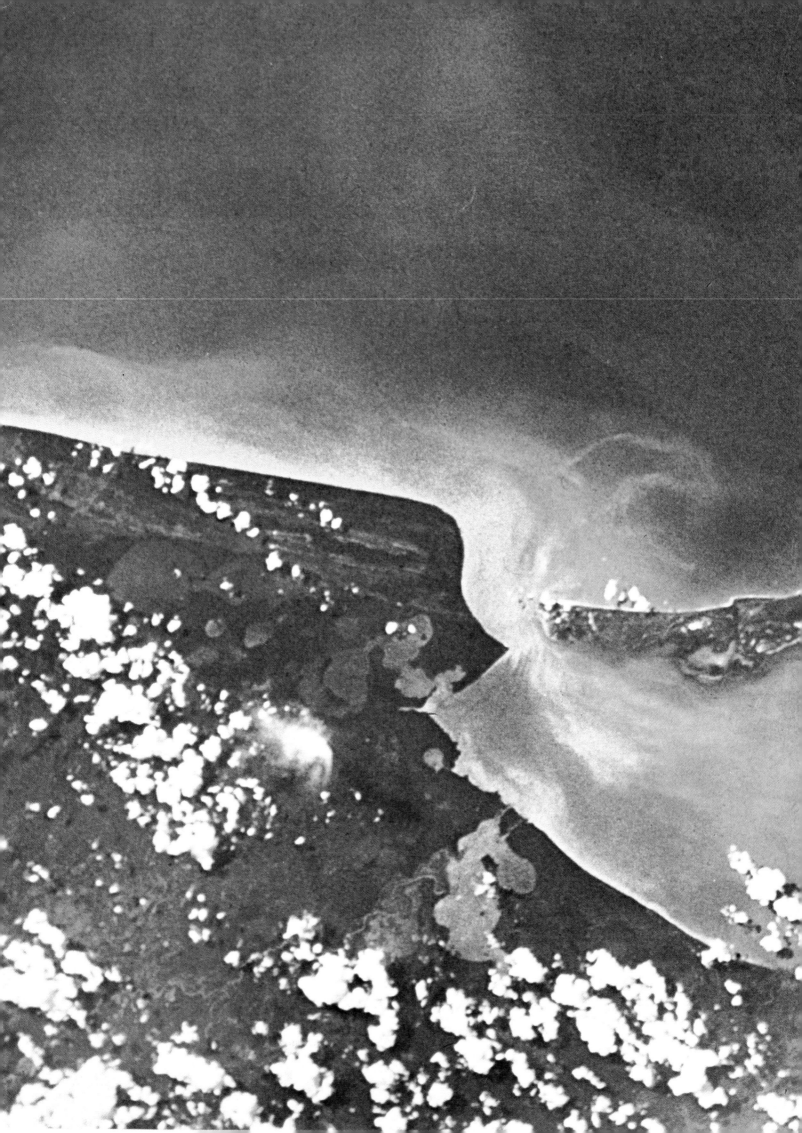

result of the larger areas of land in the Northern Hemisphere's tropical zone. As the overhead sun 'moves' northward in May and June, these land areas heat up more quickly than the marine areas and tend to pull the 'thermal equator', which marks the zone of most efficient heating, further north. In November and December, when the overhead sun moves southward in the Southern Hemisphere, the large ocean surfaces warm slowly so that the convergence zone is not displaced as far to the south of the equator as it had been to the north.

Climatic Conditions

In general, seasonal differences in weather become greater poleward from the equator, with the largest changes experienced by areas which have their weather in winter influenced by the descending air of the subtropical anticyclones and weather in summer produced by the converging air of the equatorial trough. These regions will experience an annual cycle of warm, dry winters and warm summers with heavy rainfall, the hottest temperatures occurring just before the summer rains when the increased cloudiness reduces the daily solar radiation received at the ground. On such a basic rhythm are superimposed the shorter term changes of weather associated with the trade wind and convergence zone disturbances mentioned before; the essential effect of these weather systems is to bring shorter periods (up to a week) of increased cloudiness, rainfall and wind changes.

Increased distance to the poleward margins of the intertropics eventually results in aridity, especially where large areas of land are found at latitudes $23\frac{1}{2}°$N. and S. The North African Sahara and the Middle-Eastern deserts are areas where winters are dry under anticyclonic conditions and where summers are dry because inflowing air has lost its moisture by convection rainfall over the land it has traversed *en route*.

Despite these factors there are still disturbances which can, very occasionally, bring rain to the desert.

(Far left) these towering cumulus clouds have been formed by warm, moist air encountering hills or mountains. Condensation in the displaced air releases latent heat which keeps it warm enough to 'bubble' up through surrounding air. The clouds lean over as they penetrate into regions of stronger winds. As the cumulus towers are carried downwind away from the high land they are dissipated by evaporation

(Left) cumulus clouds become more and more massive at the onset of the summer, rain bearing monsoon. Most of the rainfall is produced by weather systems travelling within the monsoon current

The Monsoons

The juxtapositioning of the Indian Ocean and the huge land mass of Eurasia across latitude $23\frac{1}{2}°$N. leads to a climatic occurrence called the Asiatic monsoon. In the summer months the hot land surfaces give rise to a low-pressure trough in the lower atmosphere which is in fact the most northerly displacement of the 'thermal equator'. The air drawn into this convergence zone comes from the SE. trades of the Southern Hemisphere, drawn across the Indian Ocean, crossing the equator and finally passing across Ceylon, India, Pakistan and Burma as the south-west 'rain-bearing' monsoon. The direction of air motion has to change as the equator is crossed, and the air current moves poleward towards the axis of rotation. Further east, in Indo-China and China, moist marine air is drawn into the continental interior across the Timor Sea and across the Pacific Ocean. Summer rainfall in the Asiatic monsoon lands is highly variable. This is because the rainfall is produced by tropical disturbances within the monsoon current. There is also a decrease in the moisture content of the atmosphere as the air penetrates towards the interior of the land mass.

The picture in January is completely reversed with the intertropical convergence now located over the Indian Ocean (just south of the equator) and extending over northern Australia. The Eurasian land mass is overlaid by a shallow anti-cyclone produced by the cooling of the land mass being communicated to the atmosphere above it, thus creating a colder, more dense mass of air at low levels. Dry, cold air tends to descend and blow outwards from the interior of the land mass as an airstream known as the north-east monsoon. This airstream changes direction on crossing the equator on its way to the convergence zone, becoming a north-westerly flow which eventually arrives in northern Australia as a moist, northerly monsoon current. This distinct reversal of winds in the lower and middle troposphere over Asia is associated with latitudinal changes in the position of the subtropical jet stream. During the winter period when the Indian subcontinent is furthest from the rising air of the intertropical convergence zone, the jet traverses the subcontinent south of the Himalayan highlands but, as the equatorial trough is set up in the Northern Hemisphere in response to the summer heating, so the jet moves to lie to the north of the highland region.

The atmosphere imposes a very strong seasonal (mansin is Arabic for season) rhythm on life in the Asiatic monsoon lands, but further interest

(Right) wispy cap of cloud over Morro Rock on the Californian coast, U.S.A., shows how even a small upward displacement of a very moist airstream can cause sufficient cooling for condensation to take place

(Below) mountain peaks on subtropical islands (like the Canary or Hawaiian Islands) may be within the dry air above the trade wind inversion, while the lower slopes are in the moist, marine air below the inversion. This photograph shows Tenerife, Canary Islands

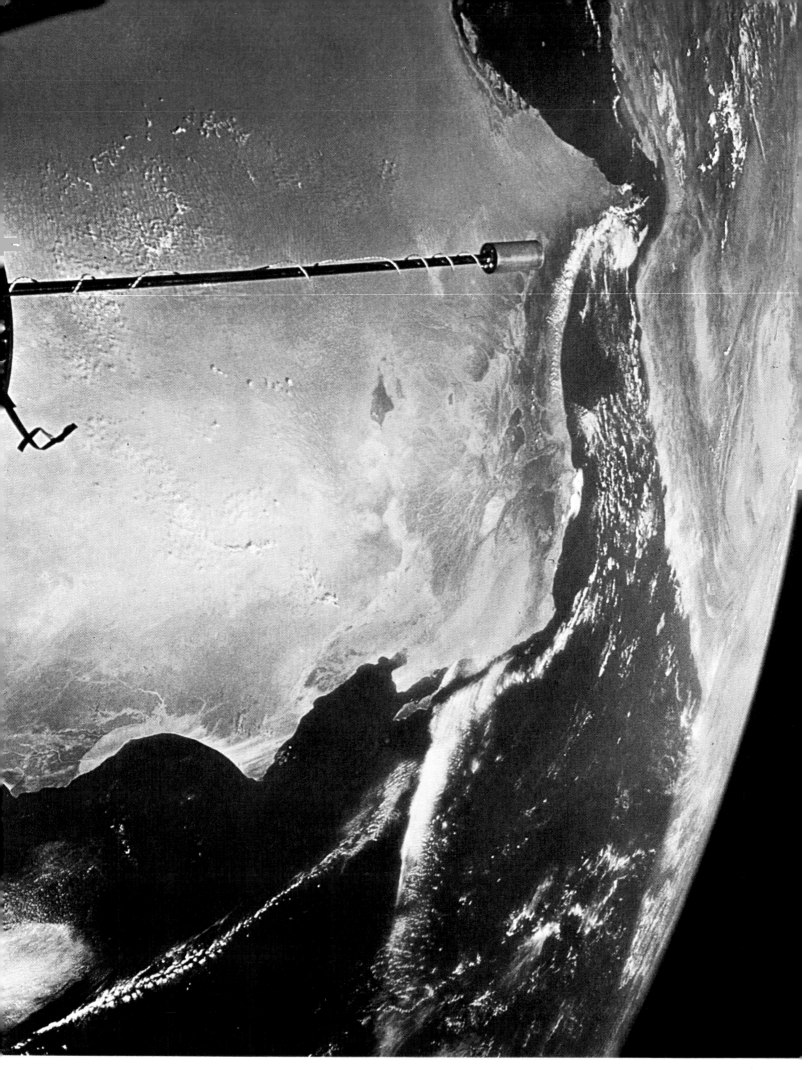

is found in this Indian Ocean area of the low latitudes for it appears that due to an accident of geography, only here do the two hemispheres exchange significant masses of air across the equator. Elsewhere the average large-scale circulation of the tropics seems to comprise independent circulation cells in each hemisphere and no exchange takes place.

Since such large areas of the intertropical belt are covered with oceans, much remains to be discovered about the meteorology of this region. Satellites have provided more questions than answers. Their cloud pictures and other data have exploded the myth of equatorial and tropical weather being regular and predictable. In this 'fire-box' zone where convective instability is fundamental to the workings of the whole atmosphere, the way the atmosphere organizes this convection and the consequences of variability in convections are both factors which still remain to be determined.

(Above) towering cumulonimbus cloud looms up behind broken stratiform clouds. The 'anvil shaped' top of this massive convective cloud is silhouetted against the evening sky and shows that the upper-level winds draw out the ice crystals which make up the top of the cloud

(Left) the warm and dry, gently descending air of the subtropical high-pressure regions leads to arid zones over land surfaces. This view, recorded by Gemini XI astronauts, is of part of the arid belt which extends from the North African Sahara desert, through Saudi Arabia and Iran, to Pakistan and north-western India. The part of this belt shown here includes the Persian Gulf, flanked by Saudi Arabia, Oman and Iran

THE MID-LATITUDE ATMOSPHERE

At about 40° latitude in both hemispheres the long-term averages of radiant energy gained and lost at the earth's surface come to a balance. Poleward from this parallel the ground loses more energy in the form of infra-red radiation than it gains from the sun in the shorter wavelengths. At latitude 70° annual incoming energy is only one half of the outgoing amount. If no mechanism for transporting energy existed the middle and high latitudes would cool down, thus decreasing the amount of outgoing radiant energy until this came to balance with the solar supply. Fortunately the oceans and atmosphere transport energy in various forms (e.g. sensible heat and latent heat) and temperatures are kept above those predicted by calculations of radiation balance.

The greatest poleward transports take place across the latitudes where subtropical anticyclones persist. These high-pressure systems mark the transition between the tropical zone, where the atmosphere gains surplus energy, and the middle and high latitudes where the air loses it. The day-to-day expressions of this cycle are found in weather and, although the fundamental reasons for atmospheric motions are known, the actual weather events can be governed by unknown minor variations – a fact which can be frustrating.

The Distribution of Pressure

The annual average pressure at sea level begins to decrease poleward of latitude 30°. Minimum values are found at about 60° latitude where extensive areas have pressure even lower than that found in the equatorial trough. In the Southern Hemisphere this low-pressure region girdles the globe to give a continuous trough off the Antarctic coast. In the Northern Hemisphere low pressure occurs at about the same latitudes but is concentrated into two oceanic regions, the North Atlantic (where a low pressure is centred roughly on Iceland) and the North Pacific (where one is found centred about the Aleutian Islands).

A quick look at a globe reveals the geographical differences which cause this assymetry between the hemispheres. Only a small proportion (just over 4 per cent) of the earth's surface south of 30°S. is occupied by land until Antarctica is reached, whereas in the Northern Hemisphere a much larger proportion (about 54 per cent) north of 30°N. is dry land and there is a sea-covered polar cap. The different rates of heating and cooling when solid and liquid surfaces are exposed to radiation make the Northern Hemisphere's separate centres of low pressure more understandable.

The Eurasian and North American land masses have higher pressure at sea level because of loss of energy from the surface which means cooling and sinking of the overlying atmosphere. In dry air density increases by just over 7 per cent as temperature decreases from 15°C to −5°C (59°F to 23°F).

The ocean surface also suffers the same annual deficit of radiation at these latitudes but, since warm water overlies colder water and drifts poleward from low latitudes, the North Atlantic and North Pacific oceans transfer heat into the atmosphere, thus creating the marine low-pressure zones. The same oceanic effects exist, of course, in the Southern Hemisphere. In the absence of continental land masses in the middle latitudes, however, the ocean currents carry heat right up to the coasts of the Antarctic. This gives rise to the trough of low pressure running through the southern oceans. Although the

(Right) the astronauts of Apollo 17 recorded this view of the earth and its atmosphere. The Southern Hemisphere dominates the picture with the Antarctic and African continents clearly visible. Spiral and arc-shaped cloud systems, associated with troughs and cyclones in the middle latitudes of the Southern Hemisphere are shown very well. Further north, note the almost cloudless areas of the subtropical high-pressure regions

oceanic transport accounts for only 25 per cent of the total amount of energy which has to be moved to balance the radiative loss, this proportion is of considerable importance to atmospheric behaviour.

The Major Wind Systems

In the lower atmosphere there must be an overall drift of air from the subtropical highs to the mid-latitude low-pressure regions. But individual air currents cannot blow directly from high to low pressure over large distances. The air is not fixed to the rotating earth so that, in the time it takes for an air parcel to travel northward out of the high-pressure region, the earth has revolved. Thus, the air appears to be deflected towards the east. This apparent deflection of the air motion, to the right in the Northern Hemisphere (to the left in the Southern Hemisphere) can be simulated roughly by pinning down the centre of a piece of paper and slowly rotating it whilst steadily drawing a straight line towards the centre. The pen or pencil travels in a straight line but the trace (the path or trajectory) will be curved to the right if the paper is moved counter-clockwise (i.e. Northern Hemisphere), to the left if the paper is moved clockwise (i.e. Southern Hemisphere). This explains the general picture of prevailing westerly winds in the mid-latitudes, since, in the lowest 3 km (nearly 2 miles) air drifts poleward to the low-pressure regions at about 60°N. and S. (there is a small polar region of easterly winds). Above 5 km (3 miles), westerly winds circle around the pole because increased height brings further poleward displacement of the low-pressure region.

Obviously, if west winds occupy most of the mid-latitudes, energy is not carried poleward in these zones by a circulation cell. Such a cell would require a change of wind direction with height as is found in the tropics (where the easterlies are overlain by high-level westerlies). The transport of energy is carried out instead by currents of air blowing side by side, south-westerly streams carrying heat and moisture poleward and north-westerly winds bringing cold air towards the equator in the Northern Hemisphere, the roles being reversed in the Southern Hemisphere. To draw an analogy it may be said that as cumulus convection in the vertical is the fundamental feature of the tropics so 'horizontal convection' plays the dominant role in middle latitudes.

The Polar Front

Air streams which have great differences in temperature and water vapour content will tend

(Above) long, drawn-out masses of cirrus indicate strong winds at upper levels and may even trace the course of a jet stream

(Left) the ice masses and the cold waters of the Arctic and the Antarctic are the sources of the cold ocean currents and cold air streams which take part in the mid-latitude weather systems

to mix, but because of the scale of the systems they do not merge their characteristics quickly. Thus, between two such wind systems, a narrow (perhaps 8 km/6 miles) boundary zone (termed a front) exists. This front moves in a manner dictated by the relative strengths of the air-streams. The broad-scale picture of winds and pressure in the middle latitudes should show the presence of one major front in the lower troposphere, known as the polar front, which, lying through the regions of lowest pressure, separates westerly air coming from the warm, low latitudes from easterly air coming from the cold, polar regions.

This general front should die away with height above 3 km (2 miles) as westerly winds dominate the atmosphere with increased height. But on a day-to-day basis one finds that often more than one major front is present and some fronts can be traced upwards for about 10 km (6 miles). Thus to explain mid-latitude weather some elaboration of the general picture is necessary.

Waves in the Westerlies
We have seen that an 'apparent' deflecting force

has to be considered when looking at bodies in motion relative to a rotating earth. This force can be shown to be related to the speed at which the earth's surface turns and therefore its strength is related to latitude since latitude is also a measure of distance from the axis of rotation. In fact, this force is at its greatest at the pole and becomes zero at the equator. Air which moves north or south across a latitude circle moves into a place with a different value of the deflecting force and tries to change its motion to balance the new force.

Well above the earth's surface with its zone of frictional drag, at a height of perhaps 5 km (3 miles) the fluid nature of the air produces an oscillation around the balance condition just as a weighted spring oscillates up and down after being pulled. Since this displaced air is embedded in the broad westerly flow of air in the mid-latitudes, its path assumes a wave-like shape as the westerly flow carries these oscillations around the globe.

In the upper troposphere such waves are a permanent feature of the atmosphere although their position, amplitude and wavelength are

(Below and right) fair weather cumulus. Many of these convection clouds show a tendency to spread outwards with fairly smooth cloud tops. This indicates that the air some distance above the surface is tending to resist the upward growth of the clouds. The cumulus clouds are associated not only with warming of the land surface but also with air currents being displaced

(Right) heat energy is transported poleward by warm ocean currents. This colour translation of an infra-red picture, taken by a meteorological satellite in February 1971, shows the Gulf Stream sweeping northward just off the south-eastern coast of the United States. Reddish-brown colours indicate surface temperatures above 20°C (68°F), blue tones about 8°C (46°F)

variable to some extent. Typically, one sees the upper troposphere in the middle latitudes as a meandering, westerly air stream with perhaps four or five wavecrests around the hemisphere. Thus a typical length for such waves might be 6,500 km (4,000 miles); hence they are called long waves.

The Jet Stream at the Polar Front

At the top of the troposphere this meandering stream of west winds contains a concentrated region of air moving at high speed. This is another jet stream, where wind speeds in excess of 120 mph may be encountered. Unlike the subtropical jet stream, this jet is rarely a continuous feature around the hemisphere and it is found at a slightly lower height, say 9 km (6 miles). Its position is related to the position of the surface front which separates cold, polar air from warm, tropical air (the north-west and south-west airstreams which converge in the low-pressure regions). Hence it is referred to as the polar front jet stream.

The turbulent air of this jet stream can constitute a hazard for aircraft but it has greater general importance because of its relationship with the weather systems of the lower troposphere, systems which are a familiar and often dominating part of our environment.

Pressure Systems

The weather systems familiar to man outside the tropics are usually known by the air pressure characteristics associated with them. A broad division is made between high-pressure and low-pressure systems (the anticyclone and the cyclone). Day-to-day weather largely depends on the presence of one of these features; the climate over a longer period of time tends to reflect the frequency of occurrence of one system, or the average number of times alternation occurs between the two types.

Generally, the closer a particular place is to the tropical areas, the more it is likely to experience anticyclones with warm, fairly dry air. Close to the poles anticyclones again will tend to dominate but with very cold, fairly dry air. Between these zones the core of the middle latitudes will have alternations between cyclones and anticyclones but the high-pressure systems may be of either the subtropical or the polar type. Summer heating usually favours the poleward extension of the warm, subtropical high-pressure belt and the displacement of the region of changeable pressure to higher latitudes. Winter brings about the extension towards the equator of areas covered by cold anticyclones with an appropriate displacement to lower latitudes of the region of variable pressure.

(Right) bank of sea fog in the Irish Sea. Air is being cooled by contact with the cold surface of the sea, producing condensation of its water vapour into fog. The land surfaces are heated by the sun thus containing the fog to the sea areas. In the distance are convection clouds over the Welsh mountains

(Below right) cirrostratus is a high cloud formed of ice crystals. Its thin, veil-like character means that its presence can go unnoticed unless haloes are observed. Cirrostratus invading the sky is a good indicator of the approach of a warm front. Upgliding air ahead of the surface front forms this high-level layer cloud

(Left) cloud structures associated with stable air. A warm front is approaching Vancouver, Canada, but the air ahead of the front is stable (i.e. it resists upward movement of individual air currents). A thick layer of altostratus ice crystal cloud covers the sky, and below it appear some lens-like clouds (altocumulus lenticularis) which are caused by waves set up in the air flow over the mountains. The smoke layer over the city is also indicative of stable air

(Pages 40–41) as a cold front sweeps past, the massive cumulus and cumulonimbus clouds give way to clearer skies in the cold air stream

(Left) very large convection clouds (in this case cumulonimbus) can be formed when cool, dry air flows over warm water. The addition of both heat and moisture causes the surface air to 'bubble' up through the mass of cool surrounding air

(Right) low clouds and unsettled weather. Under the main layer of stratocumulus clouds, shreds or fragments of cloud are being blown along. Known as stratus fractus (or 'scud') they are formed by turbulent overturning motions in volumes of air which have been almost saturated by the evaporation of falling rain

(Pages 44–5) extensive sheet of stratocumulus seen from above. The convection currents in this cloud are widespread but not strong enough to build up tall cloud masses. Stratocumulus is considered a layer cloud for this reason. It is often associated with depressions and fronts

Anticyclones

The anticyclone has high values of surface air pressure, and therefore density, because of either widespread gentle downward air motion imposed from above, or strong surface cooling which contracts the overlying atmosphere. But air, like other fluids, has to flow outwards from the high-pressure centre. As soon as this outward moving (or diverging) air current is established it begins to be affected by earth rotation. Thus in the Northern Hemisphere it is turned to the right until it is moving almost around the area of highest pressure rather than directly away from it. In the Northern Hemisphere winds follow a clockwise course around high-pressure centres. In the Southern Hemisphere deflection to the left produces a counter-clockwise circulation around anticyclones. The winds are rather light because of a widespread and gentle descent of air from the middle levels of the troposphere.

When a warm, subtropical anticyclone penetrates mid-latitude areas, it usually does so in the form of a ridge of high pressure (that is, a protuberance polewards from the subtropical latitudes). The warm airstream will still possess some descending motion. If this situation exists in winter the cooler surfaces over which it travels cools and contracts the air, encouraging the downward motion. At the same time the overall cooling condenses the water vapour steadily so that widespread cloud sheets develop, eventually producing fog and drizzle or even light rain as

water droplets grow large enough to fall to the ground. In summer the situation is very different. because warm surfaces, particularly land surfaces which gain heat quickly when exposed to summer sunshine, tend to set up convection currents which produce cumulus clouds similar to those of the trade wind zones.

If the ridge of high pressure, or anticyclone, has pushed down from high latitudes the seasonal effect is, to some extent, reversed. In winter the cold polar air moving towards the equator is subject to relatively little warming over land and cold, clear weather is the general rule. Over the ocean some warming could be expected, especially if a warm ocean current is present, and moisture will certainly enter the air as it moves over the sea surface. The decrease in density brought about by this addition of water vapour and heat can give rise to cumulus convection clouds. In summer warm land surfaces produce the same convective cells which tend to stir up the air and thus quickly modify its original cool, dry nature.

Cyclones

Air converges (i.e. blows in from all sides) into areas of low pressure but the deflection due to the earth's rotation comes into play as the air starts to move towards lower pressure, resulting in winds blowing counter-clockwise around a low-pressure centre in the Northern Hemisphere (clockwise in the Southern Hemisphere). Most low-pressure areas seen in the familiar 'plan' view plotted on a weather map have a roughly circular shape, so the air may be visualized as spiralling into the centre. Since upward movement is quite small over large horizontal distances the spiral is a very shallow one.

Obviously air which is drawn into the cyclone (or 'low') from the poleward quarters will have very different characteristics from air drawn in from the lower latitude areas. Since mixing cannot keep pace with the inflow rate 'fronts', or density boundaries, must appear. The warmer, moister air appears on the side of the low facing the equator as a wedge-shaped area, broad at first, then narrowing as in time the warm air

spirals up around the eastern side of the cyclone. Colder air penetrates around the western side but, being more dense, usually tends to spiral downwards, thus after a time cutting underneath the poleward-moving, warm air.

It is this upward motion of the moister air which creates the cloud and rain so typical of cyclonic weather. Unlike the process of cloud and rain formation in convection currents, the rate of cooling is slow as air rises along a shallow slope so the condensation takes place slowly over a wide area giving sheets or layers of cloud and belts of rain in the eastern part of the low. This cloud and rain are associated with the 'warm front', that is the advancing boundary of warm air.

It has been discovered that most of the rain which falls in mid-latitudes starts as snow formed at heights where temperatures are well below 0°C (32°F). Paradoxically, the clouds that produce the precipitation are most often made up of tiny droplets of liquid water despite the fact that the temperature may be lower than −10°C (14°F). In this kind of environment any tiny ice particle will rapidly grow into a large snowflake and fall down through the atmosphere either to melt into a raindrop before reaching the ground, or to remain as snow.

Where the colder air spirals down around the west side and pushes underneath the warm air stream a front known as the 'cold front' is formed. Here the warm air is pushed up faster creating thicker cloud layers through which cumuliform tops may appear. Rainfall at the cold front tends to be 'showery', that is more intense but of shorter duration than at the warm front. The rain forming process still involves ice particles, indeed it is more likely that snow or hail will fall to the surface at a cold front than it is at a warm front.

The final decaying stage of a cyclone is recognized when the cold air has completely undercut the warmer airstream and all that remains is a revolving mass of cool, mixing air. What initiates the development of the cyclone is not so clear. Sometimes cyclone development can be initiated by air having to pass over long, high mountain ranges, such as the Rockies and the Andes, in which case low-pressure zones will be formed in the lee of the range.

The long waves of the upper levels of the troposphere also seem to play a part. As air flows eastward and poleward towards a wave crest (or ridge) it often has a tendency to spread horizontally (or diverge). This divergence aloft can only be maintained by air rising up from lower levels which in turn creates a surface low-pressure zone

with air flowing in (convergence) to maintain the vertical flow. If the low-level inflow has a high concentration of water vapour then this will condense as the air cools on rising, adding its latent heat to the system, and thus helping to keep the circulation going. Local but large changes in speed in the upper waves help to trigger off this process and very often it is the shifting position of the polar front jet which produces such velocity changes.

The Transport of Energy

Depressions are a common feature of the mid-latitude atmosphere but each individual low-pressure system, like an eddy in a stream of water, forms, grows and then dissipates whilst being carried along in the atmosphere. During its lifetime it brings significant short-period changes in the weather to the areas it traverses, but, at the same time, it acts as an energy exchange mechanism which maintains the average temperatures in the middle and high latitudes. It brings about the poleward movement of warm air and enables cold air to move towards lower, warmer latitudes. The condensation and precipitation involved in these movements adds latent-heat energy to the atmosphere.

Since maximum average rainfall outside the tropics is found at about 50°N. and S., it is obvious that energy locked up in water vapour is transported far from the tropical belts of high evaporation. In fact about one-third of the water evaporated from the tropical oceans may move into the middle latitudes on occasion, the remaining two-thirds moving towards the equator in the tropical atmosphere. Most of the rainfall in the middle latitudes is produced by transient low-pressure systems. Geographical factors such as mountainous areas (producing extra lift and more efficient condensation), and distance from extensive, open, water surfaces (reducing moisture injection into the lower atmosphere and thus reducing rainfall), impose yet more variations on the seasonal change and the weather experienced over short periods.

The Stratosphere

As one ascends above the jet stream level and enters the mid-latitude stratosphere, the scale of the weather systems gradually becomes larger.

The lowest parts of the stratosphere are very similar to the upper troposphere but higher up, at, say, 25 km (15 miles) where radiation is of great importance, seasonal differences are very pronounced. In winter low pressure is found over the dark, polar zone and winds are everywhere

49

westerly with a long wave structure similar to, but more subdued than, that of the mid-troposphere zone 20 km (12 miles) lower down. In summer, however, when at high latitudes, un-interrupted radiation is received from the sun, a high-pressure system is found over the pole, and light, easterly winds are found around the hemisphere.

The changeover from low to high pressure with a consequent wind reversal can take place very quickly in late winter or early spring, the result of a sudden warming of a considerable depth of the high atmosphere. Meteorologists are still searching for a wholly satisfactory explanation of this phenomenon since it is apparent that the warming moves downwards through the atmosphere from levels where information is very difficult to obtain. It is not certain that such great variations in the high atmosphere have an effect on weather patterns in lower, more important levels in the middle latitudes.

Stratocumulus over hills and mountains in the evening often follows the development of cumulus clouds. The convection currents which formed the cumulus die out as the sun sets, leaving a deep layer of well mixed air. This air has sufficient water vapour to form stratocumulus when lifted by the waves set up over high land

WEATHER HAZARDS

The complex structure of life on earth has developed in the basic environment provided by the sun and the atmosphere. That does not mean that all life forms are completely adapted to their environment but rather that in the natural order of things some balance exists between a living organism and its surroundings. One of the ways that the balance can be disturbed is by a change in atmospheric conditions altering either weather in the short term or the climate over a long period. Human history is full of such occurrences as famine and flood, yet there is a tendency to regard times of great misfortune as being unusual or even undeserved. The fact is that the web of life is continually rewoven in shifting inter-dependant patterns related to variable atmospheric processes and events. Air has no conception of what is normal. The relatively few constraints which apply to the atmosphere are quite remote from the day-to-day workings of the small areas of the atmosphere which produce the weather.

So, rather than adopting an apologist's approach and saying that loss of life, destruction of property, misery and discomfort are due to abnormal weather, it is far more satisfactory to know the possible ways in which atmospheric events can affect a community by creating hazardous conditions for its members. By understanding the atmospheric part of the system it should be possible to control conditions in the community so that catastrophic interactions do not take place. After all, in Ancient Egypt the flooding of the Nile was a blessing to the agricultural society, not a disaster.

Most people regard the storm as the most hazardous feature of weather. What they mean by a storm depends largely on where they live, for it is a very general term which indicates very strong winds, rain, snow or hail (therefore cloud-covered skies) accompanied perhaps by thunder and lightning. The meteorologist could ascribe these conditions to quite different weather patterns, some being more common in certain regions than in others. But the meteorologist would find it quite difficult to say just how violent any one of these systems could be. Furthermore such information would be useful only if a community were able to relate it to the activities and structure of community life.

The hurricane is a particularly violent system which is a feature of the tropical world and some mid-latitude areas close to the tropics. It represents the most intense development of a tropical storm and, fortunately, in order that this development can take place, particular conditions are needed. Before looking at these conditions one should identify what is meant by a hurricane.

It is an intense low-pressure system, roughly circular in shape, with a diameter of perhaps 500 km (310 miles). If one could move through the hurricane measuring the air pressure at points close to the earth's surface, it would be found that the rate of fall of pressure increases towards the centre. The surface pressure at the centre might be about 5 per cent less than that at the perimeter of the system. This can only imply very rapid upcurrents of air at or near the centre so that the inflow winds must be correspondingly strong. In fact the low-pressure system is not called a hurricane unless the winds just above the earth's surface are faster than 75 mph. Very often they are above 100 mph. The strongest winds circle the system (counter-clockwise, in the Northern Hemisphere) just outside the centre. In the 'eye', or centre of the storm there is a region of light winds roughly 50 km (30 miles) in diameter.

The cloud structure too is dramatically different in the 'eye'; the spiral bands of clouds which

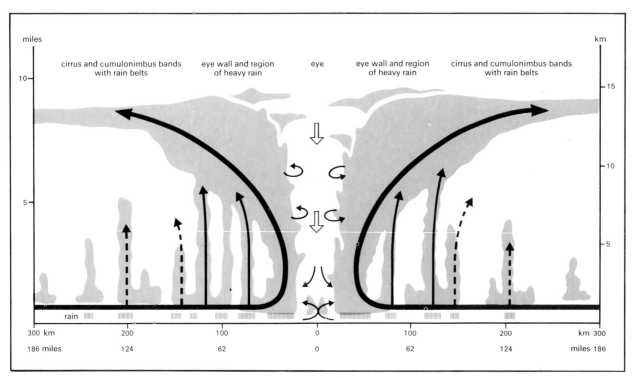

Diagram 4: idealized section through a mature hurricane
This idealized section shows the vertical motion and cloud structure in a mature hurricane. Most of the warm air rises in the eye wall (heavy arrows) with smaller amounts rising in the peripheral cloud bands (thinner and broken arrows). Warm air in the eye probably originated in the surface inflow but was carried up in the clouds forming the eye wall and then mixed inwards before descending in the eye. Some high-level (stratospheric) air also descends in the eye (outlined arrows)

Labels on diagram:

miles / km

cirrus and cumulonimbus bands with rain belts — eye wall and region of heavy rain — eye — eye wall and region of heavy rain — cirrus and cumulonimbus bands with rain belts

rain

300 km — 200 — 100 — 0 — 100 — 200 — km 300
186 miles — 124 — 62 — 0 — 62 — 124 — miles 186

converge towards the centre form a solid wall of towering cumulonimbus around the 'eye' which is almost cloud free by comparison (see diagram 4). These towering cumulonimbus clouds 'boil' up to such great heights – say 14 km (9 miles) – that the cloud droplets near the top are replaced by ice crystals. These are then blown out from the storm centre at high levels, thus forming sheets of tenuous high cloud (cirrus). This great thickness of cloud means that little light passes down through it, and such cumulonimbus clouds have a heavy grey-black appearance when seen from below.

The rapid convection of warm, moist tropical air produces (in the thick clouds) a lot of condensed water which will obviously give heavy rainfall. Hurricanes are capable of producing rain at an average rate greater than 10 mm per hour and instantaneous rates equivalent to 500 mm per hour have been recorded. In the course of their passage exceedingly high rainfall totals may be experienced. One hurricane gave more than 2000 mm of rain as it passed over a Pacific island. It is the release of energy associated with the condensation of water vapour which drives the storm. Since the hurricane depends on this water vapour, it is only over the warm tropical oceans where surface-water temperatures are 26°C (about 79°F) that this storm can develop and be maintained. The hurricane which passes over a land area loses some of its violent energy since it can no longer draw in as much water vapour to feed the convection.

Hurricanes, also known as tropical cyclones or typhoons, are not very frequent. On average about 50 are reported each year but they tend to form in the late summer to autumn when the oceans have been warmed. It has been established that they develop from disturbances in the trade winds, but, whereas there are many such disturbances each year, very few develop even into tropical storms (which have no 'eye' present and

(Left) Hurricane Gladys as seen by the crew of Apollo 7. *The great bands of cumuliform cloud and the hot towers are easily visible in the arms spiralling away from the storm centre. Closer to the eye wall, the upper-level outflow of air is marked by extensive banks of cirrus cloud. The shallow 'dome' cloud (cirrostratus) close to the hurricane's eye is a feature associated with convection when inhibited at the tropopause, in this case at an altitude of 18 km (11 miles)*

(Right) colour translation of an infrared radiation picture of a hurricane. The original grey tones of the picture obtained from the satellite have been replaced by colours which show the temperatures associated with the cloud structure of Hurricane Camille. Blues and greens indicate the cold cloud tops of the vortex and spiral bands; purples indicate the warm, virtually cloud-free waters of the Gulf of Mexico

lower wind speeds) and only about half of these storms intensify to become hurricanes. Having formed in the trade winds, hurricanes drift towards the west and then, caught in the circulation around the subtropical high-pressure areas, move poleward and eastward. They eventually appear as decaying storms in the mid-latitude westerlies. Thus it is the eastern coasts of continents in tropical latitudes which are most at risk, such as the Queensland coast of Australia, the coast of Mozambique and Malagasy in Africa, the Philippines, Japan and the coast of China, the Caribbean islands, Mexico and the south-east coast of the U.S.A., and the lands surrounding the Bay of Bengal.

At the end of October 1971 a tropical cyclone from the Bay of Bengal hit the Orissa coast of India. Heavy rains and the storm tides lashed by winds of over 100 mph caused extensive flooding in which about 10,000 people lost their lives and 1,000,000 houses were damaged. In August 1969 hurricane 'Camille' battered the Gulf coast of America and was responsible for about 300 deaths and damage estimated at $1,500,000,000. On this occasion storm tides reached up to 8 m. These examples show that, although the winds alone are strong enough to cause damage, it is the storm-lashed ocean with its high tides and fierce waves which creates the overwhelming threat to life and property in low-lying coastal lands.

Experiments are being carried out upon some hurricanes to try to reduce the storm intensity by encouraging condensation and rainfall some distance from the hurricane centre, thus reducing the amount of energy which reaches the 'eye'. Very small particles of silver iodide are released into the air to stimulate the growth of cloud droplets to form raindrops. On a world-wide scale, the most important task is to ensure that the information gained from earth satellites and weather radar is correctly used in warning systems, and that the whole way of life in high-risk areas is designed to accept the hazardous conditions with the minimum amount of long-term disruption.

To some extent a parallel to the hurricane

(Left) the relative calm in the 'eye' of a hurricane is indicated by the broken, almost 'fair weather' nature of the cloud cover in the storm's centre. This photograph of a hurricane's eye was taken from an aircraft flying above the storm

hazard can be found in the mid-latitudes; the loss of life and enormous damage suffered by the North Sea coasts of Great Britain and the Netherlands in 1953 was caused by storm tides and waves associated with a vigorous cyclone. In this part of the world the frequent passage of strong low-pressure systems is expected as a normal atmospheric event, and the interaction between sea and atmosphere can be predicted taking local conditions into account. What remains to be done to minimize the hazard is to ensure that human activity adapts to the knowledge of these effects.

The hurricane is a weather system respected and feared in many parts of the world, but it is not as violent as the tornado whose concentrated power has to be seen to be believed. Curving down from a thick, dark cumulonimbus cloud, it appears as a grey funnel, tapering off towards the ground. If and when it touches the ground, its lower portions become dark with soil and dust whipped up in the rotating column while larger pieces of trees, fences and other shattered debris fall out around the bottom of the funnel. With a noise like a train it rumbles across the land devastating a path up to a quarter of a mile wide and leaving an even wider swathe of damage. The air probably moves around in the vortex at speeds up to 500 mph and the air pressure at the centre of the system is so low as to explode houses and other structures in its path.

Tornadoes are quite obviously a great hazard but since the effect is concentrated in such a narrow path and their advance is not too fast to preclude escape, severe injuries and loss of life are not common. Although other parts of the world occasionally report tornadoes, it is in North America that they occur most frequently, particularly in the states bordering the Mississippi and its tributaries. The reason for this geographical concentration is to be found in the atmospheric conditions required to trigger this storm system.

In spring and early summer cold dry air from northern Canada may push towards the south behind a cold front. This cold air then starts

(Right) tropical storms and hurricanes can assail coastal areas with flood waves, heavy rain and violent winds. Here a tropical disturbance is passing along the coast of Mozambique

(Pages 56–7) the approaching storm cloud is often first noticed as its anvil-shaped head, made up of ice crystals, spreads out ahead of the grey bulk of the lower part of the cloud, made up of water droplets. (Cumulonimbus capillatus)

undercutting warm, moist air flowing north-eastward up the Mississippi lowlands from the Gulf of Mexico and forcing the warm air to bubble up producing convection clouds. If, at the same time, the westerly air stream is diverging at upper levels, the effect will be the same as placing a chimney over a fire. The warm air is drawn up even faster, creating large cumulonimbus clouds with turbulent air motion within them. It is from such high, turbulent storm clouds that the tornado funnel may appear, grow down to the ground, then move with the frontal cloud mass towards the north-east at about 30 mph. More than one tornado may be formed and each funnel may track across many miles before lifting and dissipating.

Because the general conditions in which tornadoes may form are easily recognized by meteorologists, warnings can be given to areas likely to be affected. Radar can then be used to survey the cloud belt and thus locate tornadoes as they form. By opening doors and windows before taking shelter in a cellar, those living in a threatened area can reduce the likelihood of injury and reduce the amount of damage done to their property.

Even so the tornado is to be feared. In January 1969, 32 people were killed and over 200 were injured when a tornado moved across part of Mississippi, U.S.A., and in February 1971, in the Mid-West tornadoes killed 108 people and caused damage which amounted to several millions of U.S. dollars.

A strong cold front, such as those of the Mississippi and Missouri valleys and the eastern United States, can be hazardous even if tornadoes do not form. The band of massive convection cloud which occurs either at, or just ahead of the front is sometimes called a 'squall line'. This name indicates the strong gusts of wind and the heavy rain or hail showers which may be experienced as the cloud belt passes over. The line is essentially made up of cumulonimbus clouds producing a band of thunderstorms.

The thunderstorm represents the fully developed stage of deep convection currents (or cells) with warm, moist air bubbling upwards and expanding, at upward speeds approaching 60 mph. In the uppermost layers of the cloud – about 9 km (6 miles) above the surface – ice crystals replace the liquid cloud droplets produced by condensation. In the turbulent air these ice crystals grow rapidly. Because the updrafts are so strong they cannot fall until they are very large, but such large crystals are much more likely to be broken up in the gusty air currents. Sometimes ice fragments are swept up and down, alternately

colliding with water droplets in the warmer – around 0°C (32°F) – regions of the cloud and then small ice crystals in the colder cloud top at –20°C (–4°F). The result of this accumulation of ice from crystals and from the freezing of water in supercooled cloud droplets is the hailstone which falls out of the cloud when it is too heavy to be held up by the upward currents.

The strength of a convective cloud largely determines the size to which hailstones grow, so those parts of the world where very warm, moist air can be lifted up to form massive thunderstorms which can even penetrate the lower stratosphere, are also places likely to suffer damage from heavy hailstorms. Exceptionally big hailstones may be over 500 g but much smaller hail can break windows, dent the roofs of parked cars, destroy greenhouses and damage or flatten crops. In June 1969 Amarillo, Texas, U.S.A., suffered a very severe hailstorm (stones of about 6 cm were seen) which caused damage estimated at $15,000,000.

Hail is very often accompanied by strong, gusty winds (up to 70 mph) which are produced when air currents blow down through the lower layers of the cloud and then spread out to blow over the ground. This is cloud air which has been dragged down as the rain or hail fell through it, and, since it cools further by evaporating some of the precipitation, the gusty outflow can be cold as well as strong. In a line of storm clouds this outflow acts as a miniature cold front pushing up warm air which will add further 'fuel' to the squall line.

In order to minimize damage to wheat fields or vineyards, attempts have been made to add silver iodide to growing cumulus clouds to produce more, but smaller hailstones, or even to 'rain out' the moisture in the cloud before it grows too large. It is difficult to say whether such experiments are completely effective but so much can be lost in a severe hailstorm that even a slim chance of success is attractive. The same silver iodide seeding technique has also been used to reduce the chance of lightning occurring in storm clouds. A fire started by lightning can cause a great deal of damage to valuable forests, and, since it is relatively easy to assess when the forest is so dry that risk of fire is great, it is sometimes economically worthwhile to seed clouds which might spark off a fire.

Lightning is a giant spark – a discharge with several thousand amperes of current flowing in a fraction of a second between the ground and the cumulonimbus cloud. The discharge occurs when the air can no longer act as an insulator between the ground, which is electrically positively

charged, and the bottom layer of cloud which is negatively charged. The top of the cloud is also positively charged which gives us an idea as to how this 'charge separation' is brought about. All cloud droplets are not the same size and as large droplets pass or bounce against small droplets, they acquire a small electrical charge by induction due to the earth's magnetic field. Myriads of droplets are present in a large cloud, so the total charge can become very large. Large drops, which acquire negative charge, tend to accumulate low down in the cloud, whilst small, positively charged drops are swept up to the top. Thus the separation of charge which produces lightning is built up within the growing cloud. Thunder is the sound shock-wave produced when the air, pierced by the discharge, is violently heated and expands with explosive force.

Tall buildings, mountain tops and even isolated trees can carry part of the ground charge closer to the bottom of the cloud. By thus locally 'compressing' the electrical field, they are more likely to be points where the discharge takes place. The lightning conductor, made of copper rod, is therefore an essential protective device for tall structures in areas where thunderstorms are expected, however infrequently.

The strong updrafts in a cumulonimbus cloud

can produce large amounts of condensed water (up to 250,000 tons at any one time). Consequently when rainfall does occur it is likely to be very heavy, though of short duration. A rate of 50 mm per hour would not be unusual in a large thunderstorm; record rates for very large storms are as high as 400 mm per hour. Obviously flooding may be a very real danger in these situations, the severity of the flood damage depending very much on local circumstances. Flash floods in desert or semi-desert areas can be produced by thunderstorms and in a matter of minutes a dry creek can become a foaming torrent. In mountainous areas with snow-covered slopes, avalanches and melting may occur if the snow is close to melting point and melt waters would be added to the already swollen mountain streams.

In lower lying land, a flood wave may progress down a river overtopping the banks and flooding farms and towns. Very often it is possible for the meteorologist to give a realistic estimate of the maximum precipitation that an area might receive from a convective storm. This enables the community to invest in such protective devices as strengthening the river bank and building holding reservoirs which, constructed along river channels, serve to delay and reduce the effect of flood waters. Above all, a warning

(Above) from bases in the south-eastern United States, weather reconnaissance aircraft not only track hurricanes, but also study their structure. These 'hurricane hunter' aircraft are equipped with instruments to measure the storm's intensity, and rely on radar to locate and track the hurricane system

(Right) tornado moving across the open landscape of Oklahoma. The dark funnel (with an inner core visible close to the top) extends from a massive storm cloud. The low pressure in the tubular vortex causes condensation, thus making the vortex visible. The explosive force where the tornado touches the ground is seen by the debris thrown up at the base. (The orange streak at the top of the picture may be a photographic fault, but no explanation is certain)

system (possibly based on radar stations able to identify dangerously large convection cells) is needed to provide time for whatever measures may be contemplated.

One should not blame the thunderstorm for all flood situations. Frontal rainfall of long duration can deliver enough water to overtax normal drainage systems and, although cumulonimbus clouds may be present in many cold fronts, it is often slow moving or stationary warm fronts which create the hazard of excessive rainfall. In this case enormous volumes of warm, moist air gently curve upward around a low-pressure centre. If the low is very slow moving a situation exists which is rather like a gigantic, inclined conveyor-belt lifting moist air and depositing condensed water from its elevated end.

Warm fronts are responsible also for what is sometimes called an ice storm or glaze storm. Such events are not common but the damage caused to power transmission lines, television antennal, telephone wires and trees can be very

(Above) pendulous elements sometimes seen on the lower edge of cumulonimbus anvils are known as mamma. They are caused by masses of ice crystals or raindrops sinking and evaporating

(Right) lightning occurs when electrical potential overcomes the insulation of the air

(Left) ships, such as trawlers, which have to navigate in high-latitude waters often face the danger of capsizing because of the accumulation of ice on rigging and super-structure. It is important that such vessels should be warned of any weather system likely to produce dangerous icing situations

severe. A thick coating of ice may accumulate on all surfaces because the rain, formed in the upgliding warm air, falls through a shallow layer of very cold air underneath. The raindrops become 'supercooled', that is they are cooled below 0°C (32°F) whilst still remaining liquid, so that as they splash onto the cold surfaces (such as power lines) they freeze and build up a coating of ice which can become so thick and heavy as to break the wires or their supporting posts. Roads resemble skating rinks; points and signalling systems on railways are put out of action; and many of the complex services on which an urban community depends can be put at risk. Ships at sea can be brought to the point of capsizing by the weight of ice on their rigging. This situation is duplicated when, in Arctic or Antarctic waters, gale force winds combined with very low tem-

peratures cause freezing spray to accumulate on the superstructures of ships faster than it can be cleared.

Again the meteorological forecast is required to give adequate warning to community emergency services, for there is nothing that can be done to change this type of atmospheric situation. Modifications are possible in the design of ships, such as trawlers, which must venture into polar seas and in this way the hazard of the ice storm may be reduced. The aircraft industry faced up to a similar situation many years ago by introducing small design modifications. Aircraft very frequently pass through layers of the atmosphere rich in supercooled water droplets which freeze onto the cold leading edges of the wings and tail plane causing a loss of control. The installation of heaters to clear away this ice has

(Right) supercooled fog, which occurs when the temperature of the fog droplets is below 0°C (32°F), can deposit a delicate tracery of ice (rime) on trees, bushes and overhead wires which can be extremely beautiful

(Pages 66–7) heavy rain showers from massive cumulonimbus clouds can be very intense and are often accompanied by cool gusty winds as air is chilled and dragged down by the falling rain

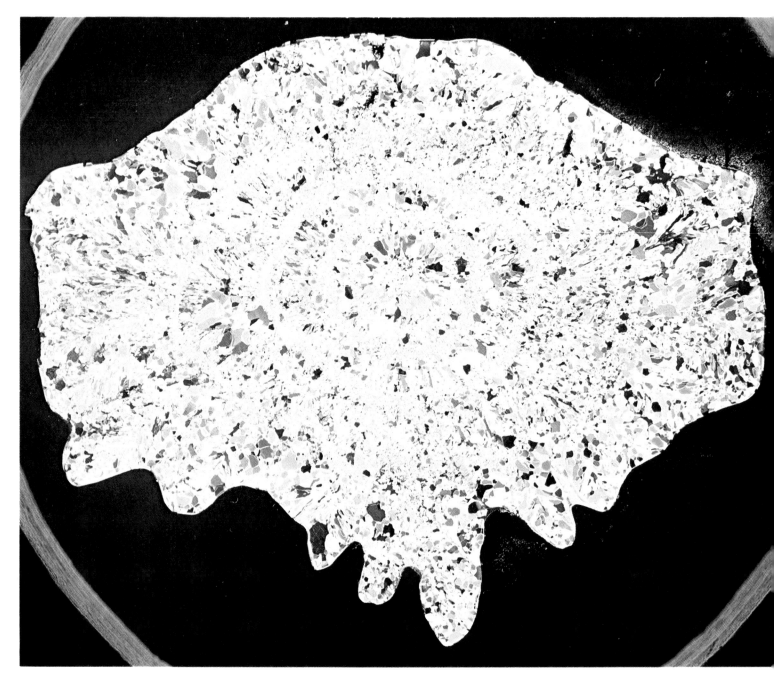

(Above) giant hailstone photographed in cross-section under polarized light. The concentric, onion skin structure is made up of layers of tiny crystals (opaque ice) and layers of larger crystals (clear ice). The opaque ice forms by the impact of supercooled droplets at very low temperatures. The clear ice develops from films of water with temperatures not far below 0°C (32°F). This particular hailstone (the Coffeyville hailstone) may be the largest ever recorded; its weight was 766 g (1.67 lb)

(Opposite, top) either rainfall in excess of the capacity of river systems or the rapid melting of snowfields can cause floods. The meteorologist and the hydrologist work together to try to establish what types of weather system create a flood threat so that a forecast of possible danger can be given

(Opposite, bottom) gradual melting of snowfields at high altitudes provides an even supply of water suitable for hydro-electricity production and agriculture. An incursion of warm air, however, perhaps with rain, can flush the melting snow out of the high valleys to create floods

been extremely successful in preventing this sort of catastrophe.

The possible dangers inherent in various types of weather are too numerous to be dealt with at length. Each part of the earth's surface has its own variety of weather conditions and each society has to learn to cope with the weather systems which endanger its way of life. Strong localized winds set up by air currents blowing over mountains, dust and sandstorms, sudden incursions of warm air which rapidly melt mountain snows and cause avalanches and floods all represent sudden changes in the environment. Man seeks knowledge about his environment to enable him to surmount these difficulties, but he must always consider the results of his own actions. A hillside stripped of forest cover exposes the soil to the destructive

force of heavy rain. Heavy outpourings of industrial smoke and fumes add to the effects of fog over wide areas. Similarly, conscious attempts to modify weather can have unexpected results which may, in some cases, lead to legal disputes. As an example of the unexpected consequences of weather modification we can look at fog clearance.

Fog is simply cloud at ground level, and is formed of water droplets small enough (under 10 μm in diameter) to be suspended in the air. The cooling of moist air which produced this condensation may have been caused either by night-time radiation cooling the ground and overlying air, or by warm air blowing over a cold land or sea surface. Another possible cause is the mixing of warm and cold air streams; but whatever the cause, visibility is greatly reduced. In fact fog conditions are defined as those which reduce visibility to less than 1 km. For this reason fog can cause multiple collisions on roads and disruption of airline services. During World War II expensive fog dispersal systems were used in which fuel oil was burnt in large quantities to stir up the air close to the ground in order to evaporate the droplets by mixing them with the air above the fog layer. Few attempts are made now to disperse so-called warm fog, when temperatures are at or above 0°C (32°F). However, if the temperatures are low and the fog is composed of supercooled droplets, then seeding with silver iodide or spraying with liquid propane may encourage the formation of ice crystals. Not long ago this technique was applied at a fog-bound Scandinavian airport. A 'hole' was cleared in the fog, but a nearby ski-jumping competition was severely interrupted by the snowfall which the fog clearance produced!

OBSERVING AND FORECASTING THE WEATHER

Radar is now being used to identify storms and cloud systems, to gain information about the way in which clouds and rain are formed in the atmosphere, and even to measure the amount of rain falling over an area

'If Candlemas be fair and clear, there'll be two winters in one year' . . . 'rain at seven, fine before eleven' – these, along with many other traditional sayings, are examples of what is often called 'weather lore' and in these little epigrams are condensed the weather experience of generations of people who observed the general quality of the weather and its sequences. Modern scientific understanding of the atmosphere requires more exact descriptions of the state of the atmosphere. So today we observe the weather with instruments capable of measuring particular attributes of the air with great accuracy. Moreover, since the measurements are given in numerical values they are not easily misinterpreted even if instrument readings from one country are used in several others.

A meteorologist who has to produce a weather forecast has to start by getting a good idea of what the present weather is like over quite a large area. Maps are the most convenient way of assembling such information but no map in two dimensions can hope to represent a three dimensional atmosphere where changes of pattern with height are of great importance. Therefore a whole series of maps of the same area is needed, starting at the earth's surface, with each successive map showing what is happening at higher and higher levels. But since the atmosphere is constantly in motion, the maps must be based on information gathered at the same time over the wide areas involved. Only thus, by building up a picture of what the weather patterns were like at one time can the forecaster use scientific knowledge to predict what will happen in the near future. In order to fulfil this requirement a vast world-wide observing system exists served by an efficient global communications network ready to flash the observed data around the world.

The most numerous observing stations are those which report the weather conditions in the lower atmosphere. It was not too difficult to set up a network of such stations in the technically advanced countries since the instrumental requirements for surface observations are quite modest. Extending the surface network to include the remote and often inhospitable parts of the world has been more difficult, although this is now partially overcome by automatic weather stations. These stations can be placed on buoys to cope with the even greater problem of finding out what is happening in the air over the oceans.

Automated devices are useful substitutes but cannot supply the meteorologist with some of the information trained observers can report. At fixed times (06.00, 12.00, 18.00, 24.00 hours G.M.T.) weather observers in many different countries measure air pressure, air temperatures, cloud cover, wind speed and direction, visibility, and the amount of precipitation since the last observation. They also note the types of cloud which are present and either measure or estimate the altitude of the base of significant cloud layers. The amount of moisture in the atmosphere close to the ground is determined and the type of weather being experienced is noted. All this information is then sent off in an internationally agreed code either by telephone, teletype or radio to a collecting centre. This centre, very often a national or regional meteorological office, transmits to other centres (probably in other countries) the information it has received from observers. In turn, it receives from those other centres their collected information. All this collection and transmission has to take place according to a strict schedule so that the information about weather over large areas can be

disseminated as quickly as possible.

The map which is drawn from this data is based on the distribution of air pressure. Of course air pressure decreases quickly with height, so that a station on a hill will always measure lower pressure than a nearby station at sea level. To overcome this difficulty all observers calculate an equivalent sea level pressure and report only this value. When all the pressure values are plotted on a map the meteorologist can then draw in lines joining places with equal pressure without worrying about the presence of high and low lying areas.

Air pressure is measured in units called millibars. The average sea level value is 1,013·2 millibars (equivalent to a column of mercury 760 mm high), but low-pressure systems typically have values less than 1,000 mbs and highs (anticyclones) have pressures greater than 1,015 mbs. The equal pressure lines, which are interpreted in just the same way as contours on a topographic map, are called *isobars*.

Each observing station has the rest of its information plotted at the appropriate point on the map and, since the map is used by different

(Above) array of meteorological instruments and measuring devices set out for long term tests. The grey tanks in the foreground measure evaporation, beyond them are rain and snow gauges and thermometer shelters

(Opposite, top) instruments which measure solar energy and the duration of sunlight are not found at all meteorological stations. Nevertheless they are of importance in some meteorological studies. The instruments nearest in this picture measure solar radiation, one being inverted to measure the amount reflected by the ground

(Opposite, bottom) observatory at Kew, one of the oldest and most famous meteorological observing stations in the world. Observations of some kind have been made here since 1773 but not, unfortunately, a continuous series

people in the course of the forecasting routine, the information is plotted according to an agreed and universal scheme. For example, the wind direction and speed at the station are shown by means of an arrow whose head is an open circle corresponding to the station location, and whose tail has a number of barbs (small oblique strokes like flight feathers) which indicate the wind speed. The direction of this arrow shows the direction of the wind (the arrow 'flies' with the wind). Symbols are used to represent cloud types, rain or snow, fog and so on. Two temperatures are shown, the air temperature and the temperature to which the air would have to be cooled to start condensation of its water vapour. This latter temperature (dew-point temperature) is a good indication of the amount of moisture which is present in the atmosphere.

A map with all this information, plotted for hundreds of stations and with the isobars drawn in, has a daunting appearance. A person with training and experience, however, can assimilate all the important features quite quickly. Of special interest are fronts whose positions at ground level can be drawn in after looking carefully at the temperatures, dew-points, clouds and winds to see where contrasting air streams are in contact. The analysis of the situation may be completed by shading in fog-covered regions, or areas where rain is falling as a further guide to rapid evaluation.

On this so-called 'surface analysis' based on the data collected over wide areas at the same time, an overall (or synoptic) view of weather systems can be seen. The next task for the forecaster is to produce a similar, though less detailed, map of atmospheric features which will be present over the same area at some future time. To do this it is necessary to consider two important things: how far the systems already depicted will have moved during the period of the forecast, and in what way the systems will change as they move. Just to add to the difficulty it is also necessary to remember that a change in the system will probably mean that its rate of motion will also change. It is at this stage that the meteorologist generally needs information from higher levels in the atmosphere. In a mid-latitude area the long waves of the middle and upper troposphere will influence the direction and speed of motion of features lower down in the atmosphere, and they may also favour new developments.

The maps for higher levels are based on information transmitted back to the ground from instruments (called radiosondes) carried up through the atmosphere by hydrogen or helium filled balloons. These sounding units measure

pressure, temperature and moisture content in sequence as they rise. The wind speed and direction is obtained from the direction and speed of their drift away from the point of release. Accurate tracking of the radiosonde can be accomplished by using radar if a reflector made from a light metal mesh is attached to the balloon.

The expense of maintaining an upper air sounding programme means that the radiosonde station network is far less dense than that maintained for surface observations. On some continents such stations may be 200 km (124 miles) apart, but over the oceans and islands weather ships provide only a skeleton network. This means that the tropical zone and the Southern Hemisphere are areas where the meteorologist has great difficulty in getting a good picture of upper atmospheric systems. The fact that the large scale features of the upper air change less rapidly than the smaller systems close to the surface makes the data coverage problem a little less severe and modern technology is being increasingly employed to improve the situation.

Balloons can be designed to rise to a certain pressure level and then maintain that level while being drifted along by winds. Such balloons, known as GHOSTs (from Global Horizontal Sounding Technique) have already been very successfully used in the Southern Hemisphere and it is now planned to have large balloons, floating at about the 25 km (15 mile) level. These will carry a supply of radiosonde units which can be released on radio command to float down through the atmosphere attached to parachutes measuring as they descend.

Satellites orbiting the earth are now making measurements of infra-red radiation flowing up through the atmosphere. From this information the temperature and moisture content at various heights can be obtained. Thus, in a short period of time, one instrument can survey around the globe, covering continent and ocean alike. In addition such satellites can locate and receive data from GHOST balloons, passing on the information to receiving stations on the ground. Other satellites 'parked' at very great heights over one part of the world (geo-stationary satellites) can not only photograph the clouds but can also, because the clouds move beneath them, provide information about the air currents which cause the clouds' motion.

As soon as the upper-air maps showing the pressure and wind systems have been drawn, the forecaster then has to use them with the surface analysis to build up in his mind a three-dimensional picture of the atmosphere. At this stage he

(Top and centre, left) any large, modern meteorological analysis and forecasting centre places great reliance on the digital computer. These machines now analyse present weather patterns and, from the analysis, produce forecasts of conditions up to three days ahead. The photographs show the computer hall of the Richardson Wing, Meteorological Office, Bracknell, U.K., and the IBM 360/195 computer

(Centre, right) hygrothermograph measures and records air temperature and moisture content on a rotating chart drum

(Bottom, left) on the chart drum of a barograph a pen traces out the changing values of air pressure. The metal corrugated drum is the pressure-sensitive element of the instrument

(Bottom, right) thermometer shelter used at weather stations, designed to stop sunlight striking thermometers whilst allowing a free circulation of air past these instruments. Four thermometers are usually placed in such a screened box (mounted on a stand about 1·5 m above the ground). They measure air temperature and wet bulb temperature (moisture content) and they record maximum and minimum temperatures

(Pages 78–9) meteorological information from remote areas can now be obtained by using automatic weather stations. The measurements made by these stations can either be stored on magnetic tape for periodic collection or transmitted directly to a manned station

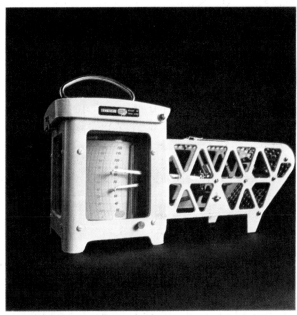

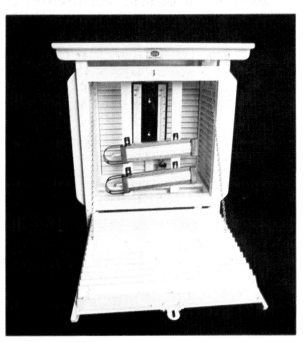

may look at the information from individual radiosonde stations which has been plotted so as to show the change of temperature and water vapour content with height. The whole process resembles a much more complicated version of three-dimensional noughts and crosses! Knowledge and experience are needed to predict the motion and changes of the weather patterns revealed by the analyses. The predictions take the form of maps (charts) for the surface and upper levels. These maps show only the pressure patterns and fronts. The forecast (or prognostic) charts can then be interpreted for whatever purpose is required. They may be used to predict the flight conditions likely to be experienced by

often best be supplied by regional forecast centres after the major prognostic charts have been produced and transmitted by a national centre.

The most significant new techniques which the meteorologist has at his disposal are 'objective analysis' and 'numerical modelling'. The digital computer is the tool which makes these techniques possible. In objective analysis a computer accepts information sent in from the observing network, checks the data for errors and uses the accepted information to produce maps of the pressure and wind systems. The advantages which are gained by using the computer are those of speed and consistency. The computer can then go on to use the pressure and motion maps

Meteorological satellites, such as ESSA 3 *(left) or a* Cosmos *satellite (right) are designed to orbit the earth at heights of 700–1200 km (450–750 miles). Pictures of cloud systems, sea ice cover and so on are transmitted back to the earth either automatically or after a command signal has been received from a tracking station*

aircraft and thus to assist in planning routes and flight levels or they may be used for public weather forecasts. Whatever the intended use, the meteorologist must assess in detail the weather conditions to be expected from the general pressure, wind and frontal patterns.

At this stage it is very often necessary to add local knowledge. Similar pressure systems or fronts may produce different weather in different locations. A front will generally produce more rain over hills or mountains than over flat plains because of the added lift given to the moist air as it passes over high ground. Streams of very cold, dry air which cross large lakes may give snow showers on the downwind shore. There may be a problem in deciding whether fog will develop in hollows and river valleys. The local detail can

together with temperature and moisture values to produce a forecast chart. To do this the computer overlays a series of grid networks, like sheets of graph paper stacked over the area, and for each grid intersection point solves perhaps six mathematical equations which are formulations of the physical processes at work in the atmosphere. Each time it has performed its calculations over the whole area it has produced a forecast for about 30 minutes. The modern high speed electronic computer works so fast, however, that it is possible to keep on forecasting in these steps until a forecast for 24 hours is obtained, and the map can then be drawn out by the machine. Some meteorological computers deal with areas as big as a hemisphere, whilst others use a global network. In this latter case the network has points corresponding to roughly every 5° of latitude and longitude in each of six layers representing height increments of 3 km (2 miles).

Not all the processes in the atmosphere can at present be represented properly by mathematical equations and small errors creep into the forecast as a result of this. These errors grow as the computer extends the forecast time. For this reason the seven-day forecast remains a difficult problem. The latest development in numerical models which predict atmospheric behaviour are experiments designed to feed into the computer information from satellites which comes in after the computer has started its calculation routine. This will be like running the computer on a production line basis rather than, as at present, stopping and restarting for each new forecast.

The long range forecast, which is an attempt to predict some of the major weather characteristics for perhaps a month ahead is still a rather experimental feature of modern meteorology. The computer routine, used for short periods, cannot be applied to this problem because of the error magnification problem, so there is a tendency to rely on the behaviour of the atmosphere in the past. Weather maps for periods of perhaps half a century may be consulted to find situations basically similar to those in evidence at the beginning of the forecast

(Below) wind speed and direction are usually obtained from instruments mounted about ten metres above the ground in a location well removed from trees and buildings. Anemometers which measure wind speed may be of the revolving cup type or propeller. Both anemometers and wind-vanes have their output signals transmitted indoors by wire leads to display dials and recorders (anemographs), like the one shown (below, right)

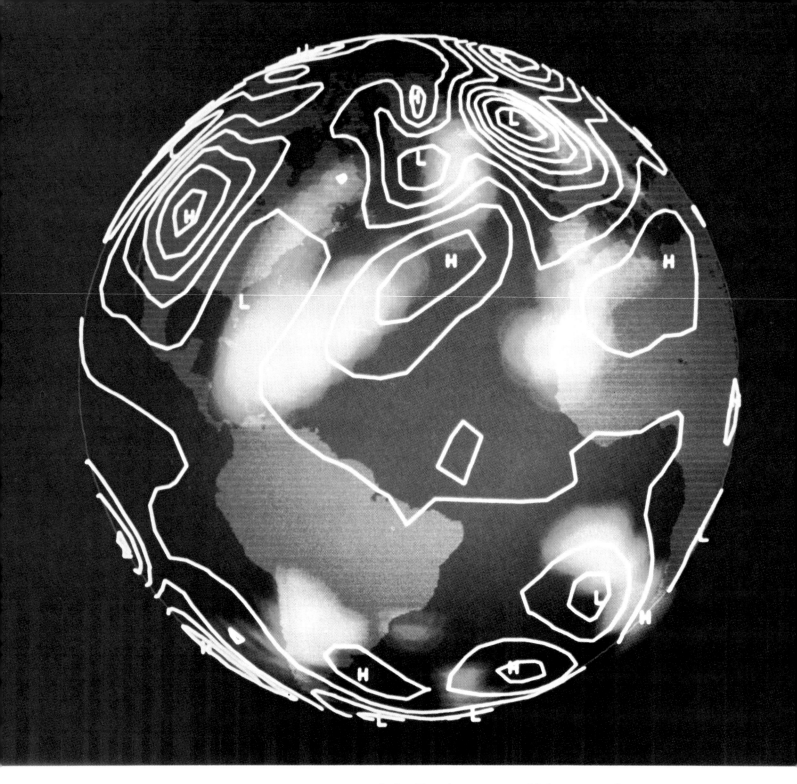

This global map of sea level pressure and cloud cover was calculated and drawn by a computer. Developing mathematical models of the global atmospheric circulation and testing them on large computers provides information which is of use in improving forecasts

period. Developments which occurred previously are studied to try to estimate the likelihood of the atmosphere repeating its past behaviour.

Some help may be obtained from the temperature distribution of the surface waters of the oceans since it has been found that water temperatures in certain oceanic areas seem to be related to basic weather patterns. For example three dry years, 1962–6, in north-eastern U.S.A. were associated with colder than normal water off the Atlantic coast of North America. This relationship seems to underline the interacting role of atmosphere and ocean which will have to be more completely understood if man is to gain more knowledge of his environment.

AIR POLLUTION

Air pollution is as old as the hills. Even before man appeared as the dominant life form on the planet the atmosphere carried some pollutants; so the most difficult question to answer satisfactorily is 'what is air pollution?' Many people think that man is the cause of air pollution and therefore a pollutant is something added to the atmosphere by human activity which has a deleterious effect on life. The implication of this answer is that outpourings of gas and ash from volcanoes, smoke from forest fires, pollens and organic compounds from vegetation, even ozone and oxides of nitrogen from lightning strokes, represent natural additions to the atmosphere. Therefore, even though they may be harmful, these pollutants are a part of the physical environment in which mankind has developed.

The reason for present concern about environmental pollution is that, whereas in the past man added fractionally to the 'natural' pollution of the environment, population growth, industrialization and modern agricultural practices have resulted in gross emissions significantly above natural levels. One very large industrial country may be injecting more than 150,000,000 tons of gases, smoke and dust into the air each year. Nor does this take into account 'thermal pollution', the extra heat added to the atmosphere where industrial cities create 'islands' of heated air.

Concern about pollution has led in some countries to legislation in which 'clean air' may be mentioned by name. The difficulties involved in such legislation are enormous; perhaps if we were logical we would only be allowed to smoke in our own homes since in enclosed public places burning tobacco produces significant pollution of the air. The meteorologist cannot decide what level of clean air is necessary for a community; he can only attempt to show what happens to pollutants in the atmosphere. The results of his investigations may indicate how a compromise can be established between the material needs of different sections of a community and the continued welfare of all its members.

One cannot proceed far in a discussion of pollution without encountering the word concentration. The total amount of air in the atmosphere is vast; even if one considers only the amount of air in the troposphere (since this is where man lives and emits virtually all his waste products) one is dealing with a global layer of air some 12 km ($7\frac{1}{2}$ miles) deep. If all emissions could be dispersed evenly through this great bulk of air then the concentration of pollutants would be very small indeed, but such efficient diffusion is impossible. Because the gases and particles tend to be injected into air layers close to the ground in localized areas, the quantity of pollutants in a given 'parcel' of air can be quite large in places, and small in others. Furthermore, the atmosphere has varied weather patterns some of which stir up large depths of air, thus promoting dispersion, while others leave the air almost motionless to absorb more and more emissions until alarming concentrations build up.

Certain rather arbitrary categories of pollutants are recognized by those who are professionally concerned with this subject. A basic differentiation is made between particles (either solid or liquid) and gases. The former are classified according to their size so that particles with diameters less than 1 μm are called aerosols, those between 1 μm and 5 μm are known as smoke and fume, while those larger than 5 μm are generally called dust. Dust particles larger than 10 μm fall out of the atmosphere quite rapidly, while those below 10 μm may be carried in the air for a number of days. Dust from the

(Left) atmospheric testing of nuclear devices contributes substantially to pollution of the atmosphere, often injecting radioactive material into the stratosphere. This photograph shows a British test made in Australia during 1956. The criss-cross pattern of smoke trails is produced by small rockets; the trails enable the wind structure and diffusion of material to be determined

(Right) growth of large urban-industrial areas like Tokyo, shown here, has created serious atmospheric pollution. Some types of weather inhibit the dispersal of waste gases and particles and this can lead to hazardous pollution concentrations. With the increase in industrial activity, the attendant increase in pollution may even cause some change in climate

Sahara is carried right across the Atlantic Ocean towards South America in appreciable quantities; 'fly ash' from industrial sources in eastern North America has been found in north-western Europe. Particulate concentration is usually measured by drawing a known volume of air through a filter pad so that the weight of material collected per volume of air can be determined. The concentration is usually expressed in micrograms (millionths of a gram) per cubic meter ($\mu g/m^3$). Averaged over a year typical urban values of particulate concentrations are of the order of a few hundred $\mu g/m^3$.

The concentrations of gaseous pollutants are expressed as a proportion of the amount of air in which they are diluted. This proportion is expressed as parts per million (ppm). The actual measurements are made by drawing known volumes of air through chemical reagents to isolate the gas under consideration. It is very difficult to establish the average length of time a gaseous pollutant may remain in the atmosphere; most gases eventually undergo chemical reactions in the air but the rate at which such reactions

take place is often determined by such things as sunshine or the amount of water vapour present.

The most common of the pollutant gases are the oxides of sulphur, carbon dioxide, carbon monoxide, the oxides of nitrogen and hydrocarbons. All these gases are injected into the atmosphere naturally, from the sea surface, from vegetation, from bacteria, and from fires or volcanic activity, but this natural 'background' concentration of these gases is quite low; for example, carbon monoxide (CO) 0.1 ppm, nitrogen dioxide (NO_2) 0.001 ppm. Human activity can produce localized concentrations much larger than these background levels. Thus an industrial conurbation may have daily average concentrations of carbon monoxide up to 20 ppm and of nitrogen dioxide approaching 0.15 ppm.

Oxides of sulphur usually enter the atmosphere as a result of burning either coal or oil, which contain sulphur. Since these fuels represent the basic energy supply for virtually all industrialized countries, the total release of sulphur into the atmosphere is very great. Each year Great Britain releases about 6,000,000 tons of sulphur dioxide and the U.S.A. about 26,000,000 tons.

Sulphur dioxide in concentrations of about 2 ppm quickly produces irritation of the eyes and throat, but since it is soluble it is not usually drawn deep into the lungs unless it is associated with aerosol particles, in which case high concentrations can cause damage to the respiratory system associated with coughing and lack of breath. Average daily concentrations of sulphur dioxide in a city usually lie between 0.2 and 0.8 ppm.

Carbon dioxide is of course a natural atmospheric gas (318 ppm in clean air) with well recognized importance, but it is also produced in large amounts in industrial cities where its concentration may be three times the background level. Carbon monoxide released into the atmosphere comes very largely from internal combustion engines and is caused by the inefficient burning of fuel. Vehicles which are accelerating or decelerating produce even higher emissions of it than a vehicle cruising steadily. Most of the exhaust fumes are released within a few feet of the ground which means that pedestrians and vehicle occupants in congested city streets are subject to high concentrations. The gas becomes lethal with continued exposure as it is absorbed

The rate at which pollutants are released into the atmosphere and the meteorological conditions in the layers of air close to the ground are both factors affecting the concentration of waste gases and particles in the air we breathe. These contrasting views of London show the effects of reduced smoke emission due to legislation and to more favourable atmospheric conditions for dispersion

into the blood stream and prevents the blood carrying oxygen around the body.

Hydrocarbons are a rather complicated group of chemical compounds produced by burning fuel and also by evaporation from light fuel oils and solvents. Some hydrocarbons are important because they react with other pollutant gases, particularly in the presence of sunlight, and irritating chemicals and ozone are formed. The Los Angeles 'photochemical smog' is perhaps the best known example of this type of reaction in the atmosphere. The main symptoms of exposure to such a pollution cocktail are irritation of the eyes, nose and throat and perhaps headaches. With the yearly increase in car population other cities have begun to experience similar pollution events.

It is worth emphasising the coexistence of these air pollutants in urban-industrial air for very often gases and particulates combine to produce more harmful effects than the sum of the individual components would suggest. This is known as *synergism*. Moreover there are many more complex substances released into the air which may be capable of harming health but whose concentrations are so low as to make both identification and effects very difficult to establish. Radioactive air pollution, that is the release of radioactive gases and particles into the atmosphere, has been the subject of a great deal of investigation since nuclear devices first introduced it to the atmosphere on a massive scale.

There has always been a background of radioactivity, produced in the earth by radioactive minerals, and in the air by cosmic rays (usually protons, the nuclei of hydrogen atoms) entering the atmosphere from space and interacting with its gases. Tritium (a radioactive isotope found in water vapour and therefore in snow and rain) and Carbon-14 are perhaps the best known forms of the atmosphere's natural radioactivity. Some radioactive pollution is released with other gases when fuel is burnt, but it is the testing of nuclear weapons and the growth of nuclear power generation which have given the most cause for concern.

Atmospheric testing of nuclear weapons creates the greatest of all pollution hazards. The test areas themselves receive the largest amounts of contamination since most of the large particles

Pollution released into the air in valleys can create a particularly hazardous environment. Air movement is constrained by the valley sides and temperature inversions often develop as cold air sinks down to the valley floor, limiting the dispersal of pollution products

89

fall out around the testing sites. Only a small percentage of the smaller particles enter the troposphere to be dispersed and eventually either deposited or washed out by rainfall over the course of a few weeks. The largest proportion of radioactive aerosols enter the stratosphere where their residence time may be longer than five years. During this time some disintegration of the radioactive isotopes takes place but many lose their radioactivity very slowly. Strontium-90 and Caesium-137 continue to be radioactive to some degree for periods of up to 300 years. Carbon-14 (produced in large amounts by weapons tests) retains some radioactivity even after 50,000 years. But all the particulate matter does not remain in the stratosphere; it is slowly leaked down into the troposphere to be carried down to the surface in rain and snow.

Nuclear reactors are of several designs and the release of radioactive gases, such as Argon-41, is very carefully controlled and checked. What is of prime concern is accidental release caused by equipment failure. This is of even greater importance since reactors are now being sited on the edges of towns and cities. An accident occurred in 1957 at Windscale on the Cumberland coast of Great Britain. The result was the release of quantities of Iodine-131, some Caesium-137 and a little Strontium-90. In a downwind corridor of agricultural land about 50 miles long and 10 miles wide, milk was unusable for months. Should such an accident happen upwind of a town it could well be necessary to evacuate thousands of people for quite a number of weeks. One can safely assume that contingency plans exist for such eventualities, however rem⌐ he possibility of

(Above) plume from this iron works shows how eddies, when they are large compared to the plume dimensions, can bodily displace sections of the plume. Downward motion may, of course, bring about high pollution concentration close to the ground

(Right) evening temperature inversion has trapped the smoke from bonfires allowing pollution to build up close to the ground

Diagram 5: behaviour of a plume of smoke from a chimney
In this diagram the generalized thermal structure of the air is shown as follows: red tones represent layers of warm air and blue tones represent layers of cold air

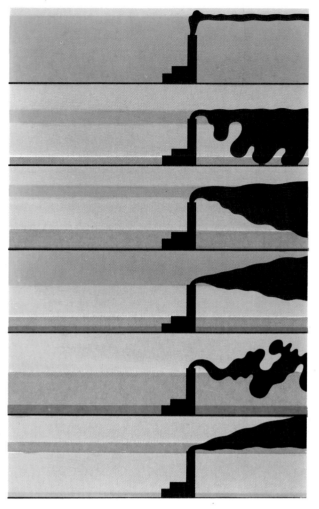

During the night, if skies are clear and winds light, warmer air may be found above the layers chilled by the ground (temperature inversion). In these conditions smoke drifts in sheets, or ribbons

In the morning the warming ground heats the overlying air and begins to break down the inversion. For a short time this stirring up of the surrounding air may bring the smoke down (fumigation)

The progressive destruction of the inversion from below (the 'lifting' of the inversion) is marked by smoke plumes becoming conical. At first, cones appear to be tilted downwards

As the normal lapse of temperature with height is set up, the cone becomes more symmetrical

When the ground is strongly heated by the sun, large, turbulent eddies due to convection currents give smoke plumes a broken-up, or 'looping' appearance

In the early evening, if the winds are light and the skies cloudless, a temperature inversion (due to radiational cooling of the ground) is established. The upward tilt of the coned plume of smoke shows that dispersion is taking place above the inversion

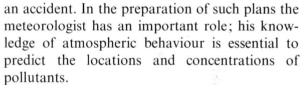

an accident. In the preparation of such plans the meteorologist has an important role; his knowledge of atmospheric behaviour is essential to predict the locations and concentrations of pollutants.

In general the dispersion of gaseous and particulate pollutants depends on the type of emission and on the wind and temperature structure of the atmosphere. The major determinant of this structure is the weather system and thus air motion and air temperature are related, but it is still possible to make an assessment of their separate roles.

Consider a chimney stack, perhaps a hundred feet tall. From its top pollutant gases and particles emerge at a certain rate but across its top the air moves (mainly horizontally) at a varying rate. When the winds are strong a larger amount of air moves past the chimney in any given period than when the winds are light. Thus, strong winds produce a smoke plume with a smaller concentration of emitted material than light winds. As the plume is carried downwind it will expand outwards because of the random motions of the gas molecules and tiny particles; the plume then takes the appearance of a cone lying downwind from the chimney which is at the apex (see diagram 5). The further from the source the lower the concentration since as each minute goes by the same amount of emitted material occupies a larger volume as the cone expands. Frequently the air is in turbulent motion; it swirls and eddies like water in a fast flowing stream. If the turbulent eddies are smaller than the plume they will tend to broaden the cone more rapidly, increasing the dilution rate. If the eddies are very large, however, they will move the plume, giving it an irregular shape but having little effect on dilution. Sometimes large eddies move parts of the plume down, increasing the concentration of pollution at ground level. So, for the most effective dispersal of emissions, strong winds and small eddies are desired, factors which will also decrease the chance of parts of the plume being brought down to breathing level.

Some general points emerge from diagram 5, the first being that, if pollutants must be released, this should be done through chimneys that are as high as possible (modern power station chimneys are often over 200 m tall) and the gases should be kept as hot as possible. Secondly, in anticyclonic conditions (high pressure, descending air, light winds) emissions should be reduced if at all possible, or cleaner fuels with a lower sulphur content used. Finally, great care must be taken in the siting of industrial plants with respect to areas of housing so that prevailing winds and local atmospheric conditions can be turned to advantage. This is of particular importance in areas with deep valleys, for cold air tends to sink down into these lower areas increasing the likelihood of periods of dangerously high pollution concentration.

In December 1930 in the Meuse Valley in Belgium and in Donora, Pa. U.S.A. in October 1948, valleys became filled with fog and pollution. Great discomfort was caused by sulphur dioxide combining with fog droplets to produce a mist of sulphuric acid capable of penetrating the lungs. Illness was widespread in both events and the death rate rose dramatically as those with chronic respiratory conditions succumbed to the polluted air. In December 1952 in London and in November – December 1962 in New York cold anticyclonic weather produced severe pollution conditions. London's 1952 'smog' (smoke and fog) lasted about five days, during which period the death rate increased enormously, particularly among those over 64 years of age. Some 4,000 people died in this pollution episode. The killer was undoubtedly the synergistic mixture of gases and aerosols trapped over the city.

Control of polluting emissions is certainly

(Below) plume from the Mt. Isa copper refinery (Queensland, Australia) illustrates the effects of the nocturnal inversion on pollution dispersal. In the early morning, before the inversion is lifted and destroyed, the warm effluent gases rise a little, but they soon cool and are then trapped by the stable air

(Right) these power
station emissions are
efficiently dispersing in
the atmosphere. The
apparently denser
plumes seen in the
reflection are caused by
the plane of polarization
of the water surface
which cuts out some of
the background light
(from the sky) in the
reflection

(Above) by using light beams as radio beams are used in radar, laser units ('lidar') can track plumes long after they have ceased to be visible

(Left) accumulation of dirt and organisms which utilize pollution products can alter the appearance of buildings. Stonework may be rotted by acids present in both the air and rainwater. The cleaning of stonework on familiar buildings like St. Paul's Cathedral, shown here, reveals the damage

possible with enough effort, but what is fundamental to community welfare at this stage is that suitable monitoring networks should exist, together with meteorological information and advice. The area occupied by the petro-chemical industry near Rotterdam is ringed with more than 30 monitoring stations. Each station automatically transmits data on an hourly basis to a control centre where meteorological information is also available. In this way areas with pollution concentrations can be combined with weather predictions to advise temporary modifications to plant operation and hence reduce the risk of unacceptable pollution levels.

In the United States experiments with computer predictions for urban area pollution concentrations are reasonably successful. Networks of monitoring stations linked to a computer are perhaps part of the answer to the problem of pollution control; another part could be played by increasing the use of devices designed to remove at least some pollutants from waste-products before emission. In this fashion we may reduce the impact modern society is making on the atmosphere, one of mankind's fundamental resources.

CLIMATIC CLASSIFICATIONS

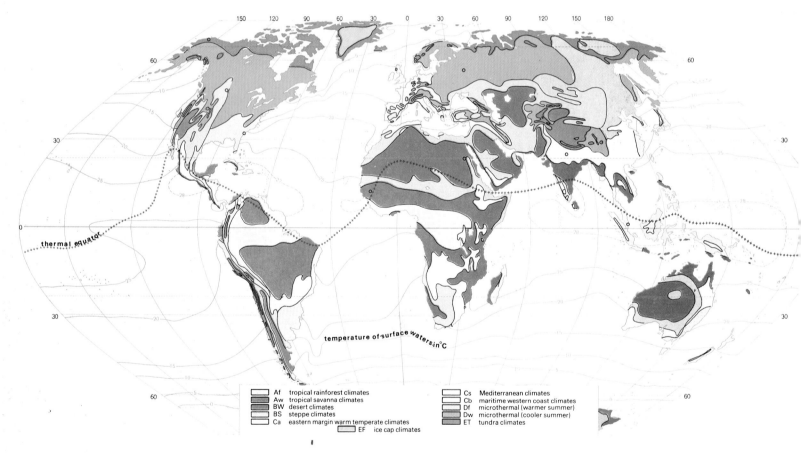

Af tropical rainforest climates
Aw tropical savanna climates
BW desert climates
BS steppe climates
Ca eastern margin warm temperate climates

Cs Mediterranean climates
Cb maritime western coast climates
Df microthermal (warmer summer)
Dw microthermal (cooler summer)
ET tundra climates

EF ice cap climates

We have seen that weather elements such as temperature, moisture content of the air, and wind speed and direction can all be accurately recorded on a day-to-day basis. After analysis, such information can be used to forecast the weather on either a short- or long-term basis. The accumulation of such data has led to the generalizations about the weather at different locations on the earth's surface which we call climate.

In places where the same factors govern weather conditions, even though such places may be thousands of miles apart they will experience similar climates. Factors such as latitude and distance from the sea will affect temperature. Areas between the tropics will receive much more effective incoming solar radiation than the middle or high latitudes. The interior regions of any large continent will always experience greater differences of temperature between summer and winter than coastal areas at the same latitude. This is because of the unequal rate of heating and cooling of land and ocean. Thus the interiors of large land

Diagram 6: Köppen's climatic classification
This map shows the climatic regions defined by Köppen, whose classification is the most widely accepted

masses will not only experience reduced amounts of precipitation on account of distance from a source of moisture, but will also have an increased range of temperature between seasons.

Another influencing factor is the position of an area in relation to the large-scale pressure systems of the world. Regions of persistent high pressure generally experience rather dry conditions at the surface as a result of the descending and outward flowing movement of air (divergence). Rain producing clouds are a feature of the permanent low-pressure belts and hence regions here will be likely to receive large amounts of precipitation. The actual amounts of rain or snow will depend on other factors such as the altitude of the land or the temperature of the ocean surface over which the air has flowed. Mountainous areas will always receive higher rainfall than lowlands if they lie in the path of rain-bearing airstreams.

annual rainfall. For this reason, the earliest known climatic classifications were made from maps of natural vegetation, since world-wide weather observations were not available until the late nineteenth century. The botanist, de Candolle, was the first to produce a classification of the world's natural vegetation in the 1860s. At the same time he was trying to find a correspondence with some of the available climatic data. He introduced the terms: *microthermal* (of low temperature), *megathermal* (of high temperature), *mesothermal* (of medium temperature) and *xerophytic* (liking dry sites) when referring to the types of vegetation found in various climatic zones.

It is only since the turn of the century that climatologists have been able to relate the average values of temperature and precipitation to their maps of vegetation distribution. The

(Above) tropical rainforest in the broad valley of the Ucayali River of eastern Peru (Af climate). It is interesting to note the oxbow lakes formed by the river

It would seem logical, therefore, to try and group areas with similar climates together. As it happens, nature has already done this for us because similar plant communities will grow under comparable climatic conditions. For a long time it has been recognized that plants exhibit a response to their environment. Thus some, like palms, cannot tolerate average temperatures much below 18°C (64°F) whereas others like many deciduous trees cannot survive without adequate moisture, needing more than 500 mm

most widely accepted scheme was devised by Köppen, one of the founders of modern climatology, who was impressed by the close correlation between the earlier vegetation maps of de Candolle and the maps of mean monthly temperature and precipitation which were becoming available. He developed a system which linked vegetation types and climatic data, also taking into account seasonal effects (see diagram 6). Initially he constructed a five-fold climatic division based on the five types of natural vegetation

(Above) rice cultivation is perhaps the most typical agricultural practice in the monsoon lands of India and Asia. This picture was taken near Madras in December, the period of the dry, north-east monsoon

Chart 1: Singapore, 1°N., 104°E. (Af climate)

	Jan.	Feb.	March	April	May	June	July	Aug.	Sept.	Oct.	Nov.	Dec.
Mean temperature	26°C	27°C	27°C	27°C	28°C	28°C	27°C	27°C	27°C	27°C	27°C	27°C
Monthly rainfall in mm	251·5	172·7	193·0	188·0	172·7	172·7	170·2	195·6	177·8	208·3	254·0	256·5

Chart 2: Cuiabá, Brazil, 16°S., 56°W. (Aw climate)

	Jan.	Feb.	March	April	May	June	July	Aug.	Sept.	Oct.	Nov.	Dec.
Mean temperature	27°C	27°C	27°C	27°C	26°C	24°C	24°C	26°C	28°C	28°C	28°C	27°C
Monthly rainfall in mm	248·9	210·8	210·8	101·6	53·3	7·6	5·1	27·9	50·8	114·3	149·9	205·7

outlined earlier by de Candolle. These divisions were further subdivided according to the seasonal nature of either temperature or rainfall, resulting in the eleven principal climatic types outlined below. It is possible to refine this basic scheme into a more complex system of subdivisions, but for general use this is rarely necessary.

KÖPPEN'S CLIMATIC CLASSIFICATION

Mean monthly temperatures are used to establish the boundaries of four of the so-called primary divisions referred to as A, C, D, and E whereas the fifth, B, is a type of climate where drought is the major characteristic; i.e. annual amounts of evaporation exceed precipitation.

A Climates (Megathermal)

These are known as the hot rainy climates where there is no month with a mean temperature below 18°C (64°F). There are two main subdivisions which are distinguished by their different seasonal rainfall:

Af – the tropical rainforest. Here the monthly rainfall amounts and the mean monthly temperatures show very little variation throughout the year. The total amount of rainfall can be very high indeed (over 2,500 mm), falling mostly during very heavy convectional showers and thunderstorms. The climate is hot and humid supporting a luxuriant type of vegetation comprising dense forests of hardwood trees such as mahogany, teak and rosewood. The dense canopy of these trees effectively prevents light penetration to the ground and for this reason undergrowth is not found except in clearings. Varieties of woody climbing plants use the trees as supports in their growth towards the light. Such vegetation types are found in the equatorial regions of the Amazon and Congo river basins of South America and Africa, and in the East Indies. (See chart 1.)

Aw – the tropical savanna. These climatic regions lie between the equator and the trade wind belts to the north and south. Constant high temperatures are maintained throughout the year but a distinct dry season is brought about by equatorward shifting of the subtropical high pressure in 'winter' whilst the poleward shifting of the equatorial trough in summer brings to these areas the associated heavy convectional rain and high humidity. But there is insufficient moisture over the year to support a dense tree cover, so extensive areas are covered by tall, coarse grass with scattered trees. Such plant cover is found in a large area of South America (mainly Brazil); a similarly large area flanks the Congo basin in Africa, and there are smaller areas in eastern India and South-East Asia. (See chart 2.)

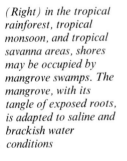

(Right) in the tropical rainforest, tropical monsoon, and tropical savanna areas, shores may be occupied by mangrove swamps. The mangrove, with its tangle of exposed roots, is adapted to saline and brackish water conditions

(Above) only drought resistant trees (such as the baobab seen here) can survive the climate of the tropical savanna regions. Much of the vegetation is made up of grassland with bushes

Chart 3: Ain Salah, Algeria, 3°E. 27°N. (BW climate)												
	Jan.	Feb.	March	April	May	June	July	Aug.	Sept.	Oct.	Nov.	Dec.
Mean temperature	13°C	16°C	20°C	25°C	29°C	35°C	37°C	36°C	33°C	27°C	19°C	14°C
Monthly rainfall in mm	2·5	2·5	0·0	0·0	0·0	0·0	0·0	2·5	0·0	0·0	5·1	2·5

Chart 4: Irgiz, U.S.S.R., 49°N., 61°E. (BS climate)												
	Jan.	Feb.	March	April	May	June	July	Aug.	Sept.	Oct.	Nov.	Dec.
Mean temperature	−16°C	−16°C	−7°C	7°C	17°C	22°C	24°C	23°C	15°C	6°C	−3°C	−12°C
Monthly rainfall in mm	15·2	7·6	12·7	17·8	20·3	22·9	15·2	10·2	12·7	12·7	10·2	17·8

B Climates (Xerophilous)

The outstanding characteristic here is one of dryness, where total evaporation over the year exceeds precipitation. Again there are two main subdivisions but in this case they are determined by the degree of dryness:

BW – the desert climates. At approximately 30°N. and S. of the equator are to be found very large cells of high pressure with subsiding dry air. Coincident with these are the large, tropical deserts of the world (e.g. the Sahara, the Arabian desert, the Sonora desert of Mexico, the Kalahari and the desert of the Australian interior). Rainfall is less than 250 mm annually, usually coming in rare heavy showers. In these tropical deserts the cloudless skies result in extremely high daytime temperatures of about 43°C (110°F) but at night strong radiative cooling takes place giving these regions their characteristically large daily range of temperature.

Desert conditions are also to be found in higher latitudes caused both by descending air on the lee side of a large range of mountains and the dry air of continental interiors. Such areas as the Gobi desert and Patagonia are as dry as the tropical desert areas, but, being much closer to the poles, the temperatures are much lower.

Vegetation in such regions is extremely sparse or even non-existent. Specially adapted plants such as the cactus and sage-brush can survive the rigours of drought and the powerful rays of the sun while other plants lie dormant in the seed, perhaps for years, waiting for the rare rainfall which will make the parched desert suddenly bloom. (See chart 3.)

BS – the steppe climate. The Steppes are the transition zones between the savanna lands and the cool deserts. The annual rainfall amounts to only 250–500 mm but much of this falls in the summer period. Since in these months temperatures are high, the loss of moisture by evaporation is also at a maximum. Trees can grow only on the banks of large rivers, but everywhere else the vegetation cover consists of short grasses such as are found in the treeless expanses of the Ukraine and Kazakhstan in the U.S.S.R., the high plains in the lee of the Rockies of North America, and in south-eastern Australia. (See chart 4.)

(Right) tropical and sub-tropical desert climatic regions (BW) have a variety of landscapes. Rock desert, sand desert, and barren mountains can all be found in these regions of aridity. Only the coarsest deep-rooted plants can survive these conditions

(Above left) Kansas (U.S.A.) prairie landscape in the mid-latitudes steppe climate (BS). Grazing and cultivation have altered the characteristics of the original prairie grasslands in North America

(Left) short grass prairie vegetation in the steppe type (BS) climatic region in South Dakota, U.S.A.

(Above) grassland in the state of Buenos Aires (Argentina). The climate is Ca, eastern margin, warm temperate, but grasslands (pampas) extend westward across Argentina into the BS, mid-latitude steppe climate found in the lee of the Andes

(Pages 104–5) the lifeless sands of the Algerian Sahara (BW)

Chart 5: Mobile, U.S.A., 31°N., 88°W. (Ca climate)												
	Jan.	Feb.	March	April	May	June	July	Aug.	Sept.	Oct.	Nov.	Dec.
Mean temperature	11°C	12°C	16°C	19°C	23°C	26°C	27°C	27°C	25°C	20°C	14°C	11°C
Monthly rainfall in mm	119·4	132·1	162·6	124·5	104·8	137·2	177·8	180·3	134·6	88·9	94·0	124·5

C Climates (Mesothermal)

These are often referred to as the temperate rainy climates, having moderate temperatures and sufficient rainfall to support some form of tree cover. Such areas are mainly influenced by the low-pressure systems of the mid-latitudes which bring much of the rainfall. The group comprises three subdivisions recognized by their distinctive seasonal characteristics:

Ca – eastern margin warm, temperate climates. The south-east of the U.S.A., most of southern China and a narrow area of south-eastern Australia experience very warm, humid summers, but mild, somewhat drier winters. These conditions are produced by the seasonal change in the prevailing winds. In summer they blow from the subtropical oceans towards the low-pressure areas over the interiors of continents. In winter the main flow of air over these south-eastern margins originates in continental high-pressure systems and is therefore much drier. This airflow is usually warmed somewhat before it reaches these regions. In the U.S.A. and Australia the proximity of the oceans keeps the winter warmer and gives appreciably more rain than falls in the interior of China. The 'natural' vegetation cover is a mixed broadleaf deciduous forest, except on light soils where extensive pine woods are frequently found. This vegetation has largely been cleared for agriculture, however, and perhaps rice and cotton are today more typical plants of this climatic zone. (See chart 5.)

Cb – the maritime west coast climate. The west coasts of the continents in mid-latitudes are dominated by the frequent passage of cyclonic storms bringing precipitation all the year round. The position of these areas, adjacent to the oceans, also results in their temperatures being moderated, so that they have a fairly small annual range and reach no extremes. Thus the summers are rather

(Below) most of the extensive areas of prairie grassland in Europe lie within the U.S.S.R., but areas of steppe grassland are to be found in Central Europe. This picture shows such an area in south-eastern Hungary

(Above) mixed woodland with azaleas and rhododendrons is found among the Great Smoky Mountains in the south-eastern United States. The humid mesothermal (Ca) climate is somewhat modified by altitude in the region of the southern Appalachian Mountains

(Right) original vegetation of the mesothermal marine climate (Cb) was deciduous and mixed forest. This has almost everywhere been cleared to make way for arable land and pasture. (River Lot, France)

Chart 6: Paris, 49°N., 2°E. (Cb climate)												
	Jan.	Feb.	March	April	May	June	July	Aug.	Sept.	Oct.	Nov.	Dec.
Mean temperature	3°C	4°C	6°C	9°C	13°C	17°C	18°C	18°C	14°C	10°C	6°C	3°C
Monthly rainfall in mm	38·1	30·5	40·6	43·2	53·3	58·4	55·9	55·9	50·8	58·4	45·7	43·2

Chart 7: Gibraltar, 36°N., 5°W. (Cs climate)												
	Jan.	Feb.	March	April	May	June	July	Aug.	Sept.	Oct.	Nov.	Dec.
Mean temperature	13°C	13°C	14°C	16°C	18°C	21°C	23°C	24°C	22°C	19°C	16°C	13°C
Monthly rainfall in mm	129·5	106·7	121·9	68·6	43·2	12·7	0·0	2·5	35·6	83·8	162·6	139·7

(Left) western slopes of the Coast Mountains, the Cascade Range and the Coast Range of the north-western United States and British Columbia, have very extensive areas of coniferous forest. This is a height modified region of the marine, humid, mesothermal type (Cb)

cool and the winters are mild.

These western margins have substantial rainfall amounts, usually with autumn or winter being the wettest period. The 'natural' vegetation found under these conditions is forest, mainly coniferous in the Americas but mixed forest is common in Western Europe. The species and stature of the trees is largely governed by the amount of rainfall (e.g. the very large douglas fir and red cedars found in areas of abundant precipitation in north-western U.S.A.). The nature of the terrain on the continental western margins determines the extent of this type of climate. In both North and South America very high mountains restrict the marine air to a coastal strip where rainfall is concentrated because the air is lifting over the mountains. In Europe, no such barrier is presented to the westerly winds and thus the influence of the ocean (by cyclonic activity) is felt much further inland. (See chart 6.)

Cs – the Mediterranean climates. Lying between the tropical deserts and the west-coast marine climates is a zone which experiences very dry, hot summers but mild, rainy winters. The annual shift of the world's pressure belts gives such

areas the climates of the two adjacent regions alternately. It has been named after the lands bordering the Mediterranean Sea where it is most extensive. It is also found, however, with cooler summers, on the western margins of the continents between latitudes 30° and 40° (e.g. the Californian coast, the Chilean coast, South Africa and the extremities of southern Australia).

Drought conditions are experienced in the summer months when the subtropical anticyclones move polewards over these areas, whereas the cyclonic activity common to mid-latitudes brings abundant winter rain. Vegetation

(Above) in the subtropical, dry summer climate (Cs) of the Mediterranean lands vegetation is adapted to dry conditions in the hot season. Drought becomes more pronounced in the southern and eastern parts of the region

is adapted to a long summer drought and takes the form of either a dry, shrubby, evergreen growth which is known as the 'chaparral' or 'maquis', or a type of woodland comprising cork oak or olive with a grassy floor. Irrigated agriculture is common, the main produce being fruit. (See chart 7.)

D Climates (Microthermal)

Unlike the three climatic types outlined so far, the microthermal climates are peculiar to the Northern Hemisphere because the Southern Hemisphere has no large continental land masses bordering the Antarctic circle. The northern boundary of these regions is the 10°C (50°F) July isotherm whereas the southern boundary is the limit of long lasting snow cover. During the winter in the D zone, snow may lie on the ground for 100 days or more, and radiation in the long hours of winter darkness results in extremely low temperatures. During the long periods of daylight (up to 18 hours at midsummer) temperatures can be high in the continental interiors, so that these regions have the world's largest annual range of temperature. This climatic region extends over much of Scandinavia, through northern Russia and Siberia, and covers large areas of Canada.

The simple twofold division of this climatic zone is based solely on the mean monthly temperatures of the summer months – the warm type having more than four months above 10°C (50°F) and the cooler between one and four months with these mean temperatures. The anticyclones of winter result in relatively small amounts of precipitation during the cold season. Summer rainfall is only a little greater because either high mountains (in North America) or extensive land areas (in Europe and Asia) prevent moist air entering these zones.

Annual precipitation is 250–500 mm but despite the modest amounts of summer moisture, extensive coniferous forests cover much of the microthermal (snow-forest) climatic area. The conifers are dormant during the winter, using the moisture from melting winter snow for growth in the summer period. Grassland occupies the warmer southern margins. In the colder regions the conifers form an open woodland with a lichen floor known as 'taiga'. (See chart 8.)

(Above) humid, microthermal (D type) climatic zones typically give rise to extensive areas of coniferous forest. Winters are cold with long periods of snow cover, hence these climatic areas are sometimes known as 'snow-forest' regions. (This picture was taken near Bayswater, New Brunswick, Canada)

(Above right) northernmost parts of the microthermal (D) climates are subarctic in character. 'Lichen-woodland', a mixture of bogs and poor coniferous forests graudally gives way to the treeless tundra further north

Chart 8: Qu'Appelle, Canada, 51°N., 104°W. (D climate)											
	Jan.	Feb.	March	April	May	June	July	Aug.	Sept.	Oct.	Nov.	Dec.
Mean temperature	−18°C	−17°C	−9°C	3°C	10°C	15°C	18°C	17°C	11°C	4°C	−6°C	−13°C
Monthly rainfall in mm	20·3	20·3	25·4	27·9	58·4	88·9	71·1	50·8	40·6	27·9	22·8	20·3

E Climates

In very high latitudes, near the polar caps, where the sun's radiation is least effective, the mean temperature of the warmest summer month is below 10°C (50°F). Precipitation amounts are low, almost all falling as snow. The 0°C (32°F) isotherm for the warmest month of the year can be used to differentiate between the zones of permanent snow and ice and those areas where snow cover disappears for a short time during the summer.

ET – the tundra. The coastal lands and islands of the Arctic Ocean and the extreme south-eastern tip of Patagonia comprise a treeless expanse of mosses, lichens and small flowering plants underlain by permanently frozen subsoil known as 'permafrost'. This is a result of the very low temperatures experienced for most of the year. During the brief period of thaw in the summer the surface layers become very wet and boggy due to the lack of drainage caused by the presence of permafrost. The short period with temperatures above 6°C (43°F) – the temperature needed by plants to grow – causes an abbreviated life cycle in flowering plants which have to flower and produce seeds within about six weeks. For the remainder of the year all plant life has to enter a prolonged dormant period under the blowing snow and extreme cold of the winter months. (See chart 9.)

EF – ice cap climate. The coldest climate on earth is found in the areas over the Antarctic ice cap and over the Greenland ice. No mean monthly temperature is above 0°C (32°F) and the South Polar region with its high altitude (about 2,800 m) has the lowest recorded surface temperatures. The extreme cold of the ice caps is due to the excessive heat loss by longwave radiation, augmented by their altitude. The Arctic Ocean has more moderate temperatures; since the Arctic ice is at sea level and is only about 4 m thick, heat can be conducted to the atmosphere from the ocean beneath. Precipitation in these regions is less than 250 mm annually so that they form cold deserts of snow and ice without any form of plant life. (See chart 10.)

(Right) perennial snow and ice cover, with pack ice on the sea, characterizes the ice cap climate (EF). This picture shows the moon setting at dawn in the South Orkneys, off the Antarctic coast

(Below) in mountainous areas the temperatures are generally lower and the precipitation higher than in surrounding lowlands. On a more local scale aspect creates climatic variations within the mountains. In the middle latitudes, for example, slopes which receive sunlight are often pastured, whilst those which because of aspect receive little sunlight have forest cover

Chart 9: Point Barrow, Alaska, 71°N., 156°W. (ET climate)

	Jan.	Feb.	March	April	May	June	July	Aug.	Sept.	Oct.	Nov.	Dec.
Mean temperature	−28°C	−25°C	−26°C	−19°C	−6°C	2°C	4°C	4°C	−1°C	−9°C	−22°C	−22°C
Monthly rainfall in mm	7·6	5·1	5·1	7·6	7·6	7·6	27·9	20·3	12·7	20·3	10·2	10·2

Chart 10: Little America, Antarctica, 78°S., 161°W. (EF climate)

	Jan.	Feb.	March	April	May	June	July	Aug.	Sept.	Oct.	Nov.	Dec.
Mean temperature	−7°C	−14°C	−22°C	−29°C	−31°C	−39°C	−39°C	−39°C	−28°C	−26°C	−17°C	−7°C
Monthly rainfall in mm	—	—	—	—	—	—	—	—	—	—	—	—

Precipitation is not given for this station as it cannot be differentiated from snow blown from the surface by the wind

CHANGING CLIMATES

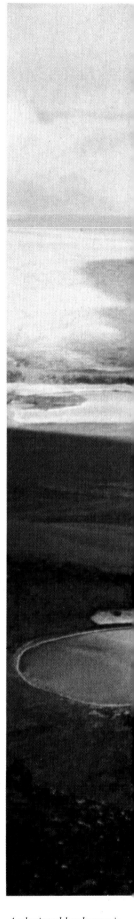

It is well known that an archaeologist digs to reveal evidence of past societies and their cultural activities. What is not so well known is that by digging deeper through the rock layers of the earth's crust one can find evidence of past climates.

Coal seams are made up of the compressed remains of a luxurious type of vegetation which must have required very warm, moist conditions. Salt deposits, now found hundreds of feet underground must have been formed when the surface of the earth at that point had a desert type of climate. Sandstone rocks preserve evidence of desert dunes.

These rocks and deposits are often present in places where the climate nowadays is quite different from that which must have existed when they were laid down. The age of the rocks can be roughly determined (coal measures of the Carboniferous period were formed almost 300,000,000 years ago) so an approximate record of climatic conditions can be carried back for millions of years.

Through these countless years the face of the earth has changed enormously: mountain ranges were built up and worn down again, the distribution of land and sea changed, and the atmosphere itself altered its composition. The closer we come to the present, the more reliable becomes the information about climates in the past, until, in the period of man's ascendancy, historical records can be consulted to give evidence of 'recent' climatic change. The important fact that emerges from all investigations is that in the long term there is no such thing as a stable climate.

The geological periods of the last few million years are of most interest because during this time it is likely the atmosphere has hardly changed in its composition. There is evidence of quite dramatic climatic changes taking place at this time, particularly in the middle latitudes of both hemispheres. If we are really to understand the atmosphere, an explanation must be found for the existence of such very different conditions to those experienced today.

An indication of the type of variations can be seen if you look at evidence from many different sources over the last 2,000,000 years. Large areas of the mid-latitudes were periodically covered by thick layers of ice and snow but some of the intervening snow free periods were warmer than the present. Climatic evidence ·gathered from both organic and inorganic remains is available. The dating and fitting together of separate pieces of evidence is a fascinating, though difficult kind of detective work.

Very large masses of ice tend to flow, albeit very slowly, and the sheer force of movement alters the appearance of the ground over which they travel. Glaciers abrade the sides and floors of the valleys down which they flow, carrying rock debris down to the point where melting occurs. Here larger rocks are dropped and the tiny rock fragments are carried away by rivers and winds to be deposited further away from the edge of the ice. Huge sheets of ice (ice domes) such as now exist in Greenland and the Antarctic, contain a large quantity of ground-up rock fragments and soil particles. Thus, when they melt, they leave behind layers of clay which contain rocks of various sizes and which may have been carried under the ice, away from their points of origin. The presence of deepened valleys, heaps of rock debris (moraines) and areas of boulder clay are therefore indicators of the presence at some past time of ice and snow masses.

Examination of the distribution of such

A glaciated landscape is being revealed by the slow melting and retreat of the Langjökull ice cap in Iceland

114

(Above) action of glaciers can be seen here forming valleys like the one shown on page 119

(Left) smooth shaped hillocks of clay containing many large rocks and small lakes or boggy hollows usually indicate that the land was once covered by a large ice-sheet. The picture shows such a landscape caused by glacial 'drift deposits' in Minnesota, U.S.A.

116

features has shown that most of north-west Europe north of 50°N. was covered by ice at some time during the last 2,000,000 years (the Pleistocene epoch) and in North America much of the continent north of 40°N. was inundated with ice at the same time. The ice sheets on these mid-latitude continents which now have temperate climates must have been at least as thick as the Antarctic ice sheet today. Thus, at times, polar climates must have occupied the middle latitudes. Similar evidence exists from the Southern Hemisphere showing that this was a world-wide effect.

Careful study of glacier and ice sheet deposits has revealed that these ice masses periodically melted away and perhaps tens of thousands of years elapsed before they were re-established. There were probably many glaciations and interglacials (the periods free of ice) during the Pleistocene epoch, but good evidence survives for only four 'ice ages' during the latter part of the epoch. To store up such large quantities of water in the form of ice on the land must have resulted in low sea levels during glacial times and high levels in interglacials. Old shore-lines can indeed be recognized, now some 120 m below the sea on the continental shelf.

The changes in temperature which accompanied the glacial – interglacial cycle can be reconstructed to a certain extent. Some extremely useful evidence is obtained from what are called 'palaeo-botanical' sources. By a process analogous to that used by the archaeologist to delve into human history, botanists can examine pollen and other plant remains preserved in the ground to ascertain the type of vegetation which flourished in the past. If one assumes that the birch trees or types of grasses which grew thousands of years ago required the same climatic conditions which they find favourable today, then the climate for the area where the remains were found can be estimated. A change of plant type with time, as revealed by changing pollen types with each succeeding layer of ground examined, gives an idea of the area's changing climate (see diagram 7).

Analysis of layers of sediments laid down on the ocean floor has also provided indications of how the temperature of the surface waters varied during climatic changes. Caribbean sediments for example show a range of about 8°C (14°F) between maximum and minimum points on the temperature-time graph. This may not sound very large but it should be remembered that enormous volumes of water must be heated or cooled before an average temperature change of even one degree could be expected throughout the oceans.

One of the difficulties encountered in reconstructing the climates of the glacial and interglacial periods can be illustrated by considering the relationships between events on land and in the oceans. Huge ice masses, such as Antarctica, are slow to respond to warming conditions, indeed it has been suggested that the delay between atmospheric warming and substantial melting of the ice mass in high latitudes may be as

Diagram 7: example of a pollen record
A pollen record, discovered from pollen preserved in peat and clay deposits, shows how vegetation has changed in the past. The major factor responsible for such changes was climate, and thus the abundance of various types of pollen can be used to identify the pattern of climatic change. This pollen diagram shows climatic conditions in Eastern Denmark from 9,000 BC to AD 500

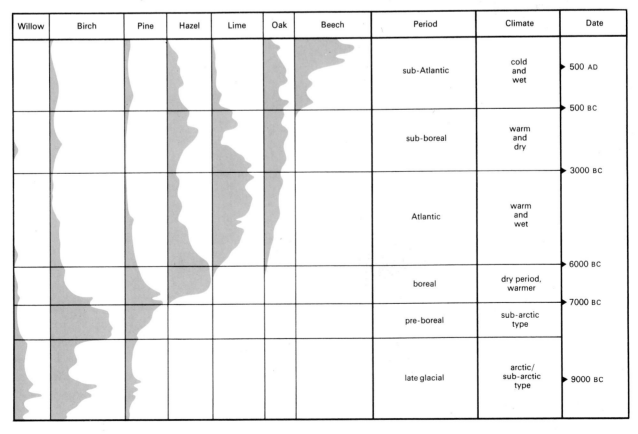

Willow	Birch	Pine	Hazel	Lime	Oak	Beech	Period	Climate	Date
							sub-Atlantic	cold and wet	500 AD
									500 BC
							sub-boreal	warm and dry	
									3000 BC
							Atlantic	warm and wet	
									6000 BC
							boreal	dry period, warmer	
									7000 BC
							pre-boreal	sub-arctic type	
							late glacial	arctic/ sub-arctic type	9000 BC

(Left) rings of wood cells produced just under the bark of a tree during the growing season can be used not only to determine the probable age of the tree but also to deduce the climatic conditions during its lifetime. Stunted trees growing in exposed places are often the best for investigation of climatic changes. Some bristlecone pines (found in California and Nevada, U.S.A.) are about 4,000 years old and therefore furnish useful records

(Right) in mountainous areas glaciers moving down valleys gouge them out into a U-shaped cross-section. Even when the glaciers have melted during warmer periods the valley shapes remain to indicate the past colder period. Additional proof of the presence of glaciers in the past is given by the moraines, ridges and banks of rocks and stones, which form at the sides and melting end of a glacier, and which are left behind after all the ice has disappeared. This picture shows the mountains of North Wales

long as 5,000 years. Such a delay mechanism must mean that the oceans warmed up before receiving the great volumes of melt water which began to raise their level.

Similarly, the long periods of time required to either accumulate or melt the great ice sheets on land may cause a discrepancy between the date of actual change in climate and the date of the evidence for this change. The end of the melting phase of the North American, Scandinavian and Siberian ice sheets of the last glacial period probably happened about 6,000 years ago, but annual air temperatures had risen to present day levels about 4,000 years before that. Problems such as this abound in the study of climatic change and until the evidence itself is established beyond doubt the task of understanding how the change occurred becomes even more difficult.

The first question to be answered is 'how did the Pleistocene epoch's cycle of glacial – inter-glacial periods start?' Evidence from rocks, in the form of fossilized magnetic fields, suggests that the continents have shifted their positions with respect to the polar axis of rotation.

Antarctica and the Arctic Ocean acquired their present polar positions perhaps about 13,000,000 years ago. The build-up of ice on Antarctica then began, and high-latitude regions in both hemispheres were probably glaciated by 10,000,000 years ago. Thus the stage was set for the Pleistocene period.

With the main continents and mountain ranges more or less as they are today one can move to the second question, 'what causes the cycle of warm and cold climates?' Unfortunately there is no one answer that can be accepted with real confidence; many possible causes can be advanced, either singly or in various combinations.

A number of years ago it was thought that only a basic change in solar radiation reaching the earth could create such dramatic changes, but there is no evidence for changes large enough to produce sufficient warming or cooling at the earth's surface. Variations in the earth's geomagnetic field can be connected with solar disturbances (sun spots) which have a frequency cycle of roughly 11 years. It may be that in high-latitude areas these can affect the weather; it seems even more doubtful as a cause of glacial periods. A much longer cycle of small solar energy variations probably exists, due to certain characteristics of the earth's orbit, but it has been calculated that temperature variations from this cause would amount to only $\pm 1\cdot5°C$ ($\pm 2\cdot7°F$) which again is considered inadequate.

Rather than looking beyond the earth for an explanation, others have sought a solution based on processes at work on the planet. The great mountain building and volcanic periods seemed to offer possible explanations. Firstly, the increase of high mountain areas in middle latitudes allowed snowfields and glaciers to form. By their very formation they could affect the weather so as to favour further snow and ice accumulation. A second possibility is that increased volcanic activity would create aerosol layers in the stratosphere thus reducing solar radiations. In 1963 Mount Agung in Indonesia had a violent eruption which, it was estimated, reduced solar radiation over the U.S.S.R. by about 5 per cent. Thus the mechanism seems plausible but, unfortunately, mountain building and volcanic episodes do not seem to have occurred with the frequency necessary to explain the glacial – interglacial cycle.

Another and more recent theory suggests that the Arctic Ocean oscillates between ice-covered and ice-free states. When free of ice it could supply moisture for heavy snowfall on bordering lands which would set off the glacial cycle. After

Volcanic eruptions eject
hot gases and particles
into the atmosphere.
Large eruptions may
inject large amounts of
dust to quite high levels
thus altering the
atmospheric effect on
solar radiation. It has
even been suggested that
large scale volcanic
activity was responsible
for the onset of glacial
periods. This picture
shows Surtsey, the new
island which has so
dramatically emerged off
the coast of Iceland

a time the cooling produced by the extensive snow and ice masses would result in the Arctic Ocean freezing over, thereby much reducing the supply of moisture needed to maintain the ice sheets. The main difficulty here is that to produce a simultaneous glacial – interglacial cycle in the Southern Hemisphere a mechanism is required whereby the atmosphere and oceans transmit the changes in some way across the equator. Since at present no one theory explains Pleistocene climatic changes some people have combined some of these theories in ingenious ways to fit the evidence, but meteorologists tend to be very cautious about putting forward theories.

The reason for caution on the part of atmospheric scientists is that they are by no means sure that the atmosphere has to work in the general manner we have observed in recent times. Since the end of the last glacial phase the palaeobotanical record in middle latitudes has shown oscillations in annual mean temperature of the order of $\pm 2°C$ ($\pm 3.5°F$). In historic times fluctuations have occurred of middle- and high-latitude temperatures and precipitation, the evidence for such variations being culled from archaeological information, tree-ring analysis, written history and, latterly, instrumental records. Relative to the geological time scale this period of so-called 'secular' climatic change is very short indeed and, as might be expected, the changes themselves have not been dramatic in the way that glacial – interglacial changes were. However, even changes as small as $1.5°C$ ($2.7°F$) in annual mean temperature have a significance for mankind.

A warm period, AD 1150–1300, allowed wheat to be grown as far north as Trondheim in Norway (about 63°N.); a cold period, AD 1550–1770, affected harvest yields and probably population numbers in north-west Europe. It is possible to reconstruct the general weather patterns which accompanied these secular changes but a cause has not yet been found. Since recent, relatively well-documented alternations in climate are unexplained how much greater is the problem of the ice ages!

Reflection on the implications to human life of recent climatic change must begin to sound a distant warning bell, for man himself is producing changes in his environment. In areas of urban sprawl 'islands' of heat are created both by the burning of fossil fuels and the added infra-red radiation blanket provided by dust and haze layers. Pollution-laden air affects visibility and reduces the amount of solar radiation reaching the ground by as much as 15 per cent. Cities tend to have more cloud than surrounding rural areas

and they may have an effect on rainfall amounts because of the increased concentrations of the small particles necessary for rain and snow formation.

When one considers the continual growth of cities and the build-up of industrial activity it becomes apparent that some small but perhaps significant changes may be taking place on a world-wide scale. Burning coal and oil has increased the carbon dioxide content of the atmosphere; since 1900 it has increased by about 12 per cent and the rate of increase continues. Carbon dioxide is an efficient absorber of long-wave radiation so that increasing its concentration in the lower atmosphere should bring about slightly warmer average temperatures. Computer calculations have shown that stratospheric cooling may also be a result of increased carbon dioxide. Some time ago the variable concentrations of the gas were thought to act as a mechanism for natural climatic change. Perhaps man unwittingly will test this theory, and, to add to the effectiveness of the 'experiment,' he is clearing forest lands – forests which would otherwise have absorbed some carbon dioxide as part of their normal respiratory process.

Particulate matter is now being spread through the atmosphere. Haziness has increased dramatically even in areas remote from industry. Between 1957 and 1967 there was a 30 per cent increase in haziness over the Mauna Loa observatory in Hawaii. Some levelling off has occurred very recently but this may not be permanent. An increase in particulates suspended in the atmosphere, as recognized by those who favoured a volcanic dust theory of climatic change, could decrease average temperatures.

A similar effect has been noticed in some parts of the world where regular aircraft passage at high levels produces condensation trails which spread out into thin cirrus clouds. It is possible that jet engines could pollute the stratosphere with water vapour which will either increase cloudiness or reduce stratospheric temperatures through radiative effects.

The first computer experiments have already been made to see whether, with our present knowledge of the workings of the atmosphere, it is possible to predict the effects of pollution on global atmospheric conditions. As yet there are very few meteorologists who would have the confidence to forecast the result of this continual, inadvertent modification of the atmosphere. The air is a fundamental part of man's environment, so much so that knowledge of the atmosphere and its workings must be advanced to allow us to guard it as a most valuable resource.

TYPES OF CLOUDS

(Below) these spectacular clouds are called wave clouds (because air rides up in waves as it passes over hills and mountains) or lenticular clouds (because in the wave crests the rising air condenses water vapour forming clouds with smooth shapes like almonds or lenses). If the air blowing across the mountains comprises some moist layers and some dry layers then the clouds too will be layered, even having the appearance of a pile of plates

(Top, left) cirrus: a high, ice crystal cloud. This type of hook-shaped cirrus (uncinus) is formed by ice crystals dropping and then being pulled out into a tail by the wind as they evaporate and grow smaller

(Top, right) cirrus and cirrocumulus: these high level clouds are made up of both water droplets and ice crystals. The droplets are 'supercooled', i.e. their temperature is well below 0°C (32°F). The rather uniform cirrus cloud is being streamed out in the direction of the wind, but the cirrocumulus is in the form of narrowly spaced billows or 'rolls' lying across the wind. The billows are formed by convection cells within the cloud layer being rolled over by winds at the top of the layer

(Centre, left) altostratus: the base is at medium levels in the troposphere but may be composed of a very thick layer of ice and snow crystals. The sun can be seen indistinctly through this sheet of cloud (translucidus)

(Centre, right) stratocumulus: low-level clouds of water droplets often associated with low-pressure systems and rainfall. This cloud can be quite thick in conditions of widespread convection, and it is often formed into rough 'rolls' by the winds

(Left) stratocumulus, evening: as night time cooling (due to infra-red radiation) begins, convection dies out and the clouds flatten, thin out and eventually dissipate

125

126

(Top, left) stratus and hill fog: stratus is a low-level, water droplet cloud which is in the form of a widespread, uniform, though often quite thin, layer. Cooling of warm, moist air as it passes over a cold region will produce stratus. The nearly saturated air close to the surface may have to rise over hills in which case condensation takes place and hill fog is formed

(Above, left) cirrostratus: sparsely distributed ice crystals in an air layer which is high up in the troposphere produce this thin, veil-like cloud. The halo around the sun is caused by the ice crystals refracting the sun's rays. Sometimes, if the ice crystals are aligned in a particular way, 'mock suns' and 'sun pillars' also appear

(Top, right) orographic cloud: an example of hill fog formed when cooling, almost saturated air is forced to rise over high ground. Often the air is trapped by a temperature inversion giving the cloud a smooth top surface. When it has passed over the high ground the air will sink and the cloud may dissipate as it does so. Much of what appears at first sight to be snow in this photograph is, in fact, cloud

(Above right) altocumulus: in this variety of altocumulus (floccus) quite strong convection currents in an air layer well above the earth's surface have produced small and irregular puffy white clouds which are made up of water droplets

(Left) altocumulus (billow clouds): convection clouds of an upper layer of air are 'rolled up' when they encounter a stronger wind higher up. Changes in wind speed or direction will affect the alignment of the billows

Selected Reading List

Barry, R.G. and Chorley, R.J. *Atmosphere, Weather and Climate*, Holt, Rinehart & Winston, New York 1970; Methuen, London 1971.

Flohn, H. *Climate and Weather*, McGraw-Hill, New York 1968; Weidenfeld & Nicolson, London 1969.

Hubert, L.F. and Lehr, P.E. *Weather Satellites*, Ginn-Blaisdell, Waltham, Mass. 1967.

Larsson, K. *Your Book of the Weather*, Faber & Faber, London 1966.

Ludlam, F.H. and Scorer, R.S. *Cloud Study: a Pictorial Guide*, John Murray, London 1957; Macmillan, New York 1958.

Mason, B.J. *Clouds, Rain and Rainmaking*. Cambridge University Press, Cambridge and New York 1962.

Meteorological Office *A Course in Elementary Meteorology*, Her Majesty's Stationery Office, London 1962.

Riehl, H. *Introduction to the Atmosphere*, McGraw-Hill, New York and London 1965, 1972.

Wicklam, P.G. *The Practice of Weather Forecasting*, Her Majesty's Stationery Office, London 1971.

Acknowledgments

Sources of photographs on the following pages:
2 top: Mount Wilson and Palomar Observatories. 2 bottom: Archivio I.G.D.A. 3 top, 98: S. Prato. 3 bottom: Sacramento Peak Observatory. 5, 6, 10, 21, 24–5, 30, 52 bottom, 53: N.A.S.A. photos. 7, 13, 50, 101, 104–5: N. Cirani. 8: Spectrum. 12: U.S.I.S. 14, 122: British Antarctic Survey. 15: photograph by G. V. Black from the Diana Wyllie filmstrip *Optical Phenomena*. 16 top, 63: I.C.P. 16 bottom, 31, 39 bottom, 56–7, 127 bottom left and bottom right: Archivio I.G.D.A./Foto Unesco. 17 top, 22, 34, 65, 69 top: T. Poggio. 18: U.S. Army photograph. 23, 35, 47: Picturepoint. 26, 44–5, 46, 125 centre left: photographs by R. S. Scorer from the Diana Wyllie filmstrip *Cloud Forms*. 27, 55: A. Filippini. 28: L. Pellegrini. 29, 39 top, 66–7, 74 top, 75, 102 bottom, 107, 108, 111, 116 bottom, 125 top left and right and centre right, 126, 127 top left: Peters. 33: Associated Press. 36, 43: Marka. 37 top: Ostumi. 37 bottom: National Oceanic and Atmospheric Administration, Washington, D.C. 38, 40–41: photographs by R. S. Scorer from the Diana Wyllie filmstrip *Cloud Recognition*. 42, 112, 125 bottom: C. M. Dixon. 48: Seff/I.G.D.A. 54: Jack Novak. 58, 60: Claude Lefèvre. 61: Frank Lane/Wayne C. Carlson. 62: Dr. J. F. R. McIlveen. 64: S. & D. McCutcheon. 68, 82: National Center for Atmospheric Research, Boulder, Colorado. 69 bottom: M. Pedone. 70: Copyright J. H. Golden, 1967. 71: P. H. Ward/Natural Science Photos. 72, 78–9: Plessey. 74 bottom: Crown copyright, reproduced with the permission of the Controller of Her Majesty's Stationery Office. 77 top and centre left: photographed by Angelo Hornak by courtesy of the Director-General of the Meterological Office. 77 centre right and bottom: photographed by Miki Slingsby by courtesy of Cassela. 80: Archivio Radaelli. 81 top: Novosti. 81 bottom: Rizzi. 84: United Kingdom Atomic Energy Authority. 85: Orion Press. 86, 87: photographs by R. S. Scorer from the Diana Wyllie filmstrip *The Air You Breathe*. 88–9, 90, 92, 93: photographs by R. S. Scorer from the Diana Wyllie filmstrip *Pollution (part 2)*. 91 bottom: photograph by R. S. Scorer from the Diana Wyllie filmstrip *Pollution (part 1)*. 94: Seeley Paget Partnership. 95: Central Electricity Generating Board, Research Dept. 97: E. Schultess. 99, 109: Archivio P2. 100: Conzett-Huber. 102 top: E. Muench. 103: S.E.F. 106: Titus/Tiophoto. 110: Information Canada Photothèque/Freeman Paterson. 113: British Antarctic Survey/ E. Mickleburgh. 115: M. E. Parker. 116 top: Swissair. 118: 1–Z Botanical Collection. 119: Aerofilms. 120–121: Icelandic Photo and Press Service. 124: British Antarctic Survey/C. Le Feuvre. 127 top right: British Antarctic Survey/Dr. R. M. Laws.
Diagram 4 (page 52): adapted from Palmén and Newton, *Atmospheric Circulation Systems*.
Diagram 7 (page 117): adapted from J. Iversen, *Naturens udvikling siden sidste istid* in *Danmarks Natur* vol. 1.